ZULU JOURNAL

ZULL

JOURNAL

FIELD NOTES OF A NATURALIST
IN SOUTH AFRICA
BY RAYMOND B. COWLES

University of California Press
Berkeley and Los Angeles
1959

UNIVERSITY OF CALIFORNIA PRESS
BERKELEY AND LOS ANGELES, CALIFORNIA
CAMBRIDGE UNIVERSITY PRESS
LONDON, ENGLAND

© 1959 BY THE REGENTS OF THE UNIVERSITY OF CALIFORNIA
LIBRARY OF CONGRESS CATALOG CARD NUMBER: 59-8760

PRINTED IN THE UNITED STATES OF AMERICA
DRAWINGS BY KENNETH KRAL
DESIGNED BY WARD RITCHIE

Dedicated to

T. R. MALTHUS 1766–1834

whose insight revealed the key

to a better future for mankind

PREFACE AND ACKNOWLEDGMENTS

This book is based on notes and collecting journals prepared in the field during two trips to South Africa, one in 1925–1927 and one in 1953, and on boyhood recollections of the early 1900's. I have not used the diary style because precise dates when observations were made are broadly irrelevant. The book is a compilation of winnowings from a large assemblage of technical notes that would be uninteresting except to a specialist. I have made no attempt to knit together closely the incidents that make up this book. But I have tried to capture the mood of the country, its seasons, and the passing years.

Geographically, the specific area in which the observations were made measures only about 200 by 100 miles, but my travels at various times sampled Africa from the Cape to Portuguese East Africa, and from Mombasa to Uganda (including the headwaters of the Nile at Jinja on Lake Victoria), and various areas elsewhere on the Continent.

I wish to express thanks to my missionary parents whose sympathy and understanding of a small boy's interest in nature, and the folklore of the native people, made it possible for me unconsciously to absorb some of the natives' point of view and their curious natural-history beliefs. Although the beliefs of my parents often conflicted with mine, they were tolerant enough to support my biological studies, thus helping to supplement the simple natural-history experience of early years and ultimately to provide the background for this book. It is also a pleasure to acknowledge my debt to my aunt, Mrs. B. N. Bridgman of (Okalweni) Kenterton, Natal, for her hospitality and encouragement.

As my mentor and guide to the natural-history lore of the natives, their language, and pharmocopeia I owe an unrepayable debt to Umditshwa of Umzumbe, whose eternal patience with a small boy contributed so much to a child's fund of information and his appreciation of the splendid character of so many of Umditshwa's people.

I am indebted to the Wenner-Gren Foundation for Anthropological Research for financial assistance that permitted a reëxamination of an already familiar terrain and made possible an appraisal of changes that have taken place in certain sections of South Africa during the past half century. Thanks are also due to the Regents of the University of California for supplementary support in the form of funds and instruments.

There are many others to whom I owe a debt of gratitude, black and white residents of South Africa, but the list is too long to make full acknowledgment and it would be unfair to list only some of those in Johannesburg, Durban, Highflats, Umzumbe, Kenterton, Pietermaritzburg, Umzinto, the Hluhluwe game reserve, Inchanga, and elsewhere, who have contributed information and hospitality: to these my thanks must go collectively.

Whatever there may be of beauty or value or prophesy in the following pages is dedicated to those who love a world unspoiled by man; and love man well enough to try to leave to future generations some unspoiled fragments of that world.

Los Angeles, January, 1959   *R. B. C.*

CONTENTS

ILLUSTRATIONS

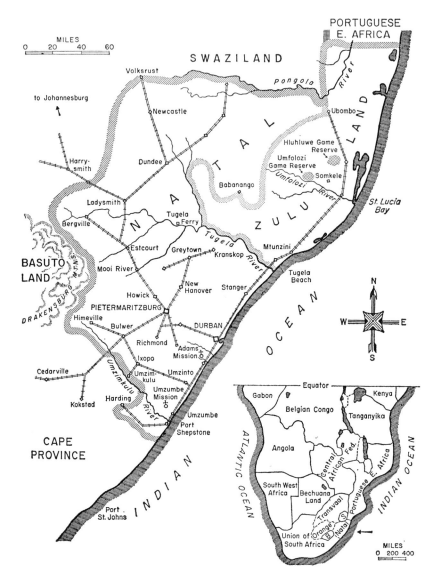

MILES
0   20   40   60

to Johannesburg

PORTUGUESE
E. AFRICA

SWAZILAND

Volksrust

Newcastle

Pongola

River

Harry-
smith

Dundee

Ubombo

Hluhluwe Game
Reserve

Umfolozi
Game Reserve

Somkele

Babanango

Umfolozi

River

St. Lucia
Bay

Ladysmith

Tugela
Ferry

Bergville

Estcourt

Greytown

Kranskop

Tugela

River

Mtunzini

Mooi River

New
Hanover

Stanger

Tugela
Beach

Howick

PIETERMARITZBURG

Himeville

Bulwer

DURBAN

Richmond

Adams
Mission

Ixopo

Cedarville

Umzim-
kulu

Umzinto

Umzimkulu

River

Kokstad

Harding

Umzumbe
Mission

Umzumbe
Port
Shepstone

CAPE
PROVINCE

Port
St. Johns

INDIAN

OCEAN

BASUTO
LAND

DRAKENSBURG MT.NS.

N A T A L

Z U L U   L A N D

N
W   E
S

Equator

Gabon

Belgian Congo

Kenya

Angola

Tanganyika

Central African Fed.

South West
Africa

Bechuana
Land

Transvaal

Portuguese E. Africa

Union of
South Africa

Orange

Natal

ATLANTIC OCEAN

INDIAN OCEAN

MILES
0   200   400

Map of the locale of Zulu Journal

INTRODUCTION

I was born early one warm, steamy morning in December, 1896. Summertime had come to Adams Mission Station, six miles inland from the South African shore of the Indian Ocean, and built near the Amanzimtoti "river"—a small creek emptying into the ocean seventeen miles south of Durban. Later I spent much time at the Umzumbe Mission Station by the Umzumbe River, eighty miles south of Adams Mission and twelve miles inland from the shore. From that station which was on an eminence I could look down on the Umzumbe River or upward to an encircling sweep of hills. Thus, from early youth, I understood the biblical lines, "I will lift up mine eyes unto the hills, from whence cometh my help."

These circumstances may help to establish the legitimacy of my life-long affection for South Africa. During this life, a man's normal span, I have witnessed biological and human changes of a magnitude both impressively good and frighteningly foreboding.

The province of Natal is a well-watered band of territory lying between the high, inland plateau broken by the uplift of the Drakensberg Mountains, and the Indian Ocean to the southeast. In topography and color, it is one of the most beautiful sections of the Continent. It lacks some of the strangeness of the, to us, exotic tropics, and yet it partakes in the strange in many ways, and this feature in combination with its pastoral domesticity, its seeming peacefulness, make it unique in its own right.

Because of the steepness of the gradient from inland to the coast one can find almost any climatic conditions. Winters are cold on the high plateau. Ice and frost are familiar winter phenomena although snow, because of the winter rainless periods, is not frequent except "on the Berg," the stupendous and beautiful Drakensberg.

Those acquainted with California will be familiar with the sharp climatic changes that can be enjoyed at different altitudes within a few miles of each other, and hence they would not be surprised by the remarkable faunal and floral changes that can be seen over comparatively short distances in the area under discussion. It was over this beautiful land that have swept since time immemorial the battles of conquest of the ancient peoples, the original occupants of this land before the Zulus and the whites came.

The observations that form the substance of this book were made almost exclusively in and around Natal and its northeastern Zululand district, where the main natural-history highlights are even more closely restricted to the native-reserve lands (broadly comparable to our Indian reservations); the intervening crown lands (somewhat akin to our national-forest lands); white men's farms; and the game reserves (in the United States often referred to as game "preserves") especially in Zululand.

The Adams and Umzumbe missions of my youth were located in the native-reserve areas, and since untilled land was abundant and bush (the low, vine-tangled South African forest, that elsewhere would be called jungle) was extensive, I lived and grew up in a naturalist's paradise. The ancient customs are of necessity breaking down now among the Zulus, but in my younger days, kraals (family villages composed of the huts of the owner and his wives and children) still were widely scattered, and each might have miles of virtually untouched grazing and bushlands surrounding them. In some instances, a man might be proprietor of several kraals for it was commonly accepted that a man could manage and keep peace with no more than ten wives in any one kraal. Because those who could afford to do so married as many wives as possible, a rich man's kraals might be scattered over an area of many square miles. I once attended the wedding of a chief when he was acquiring his hundredth wife. In this case, he had ten kraals for his convenience and the peace of his households. Villages, in the ordinary sense, were, except in and around white men's cities, almost nonexistent, and the sweep of open country between the kraals furnished some of the best and almost untouched collecting areas.

It was among such circumstances that I grew up, and it was natural that from earliest childhood I lived close to the natives and the surrounding animal life. I learned the Zulu tongue from my native nurses, female when I was very young, and male when I grew to understand more of what went on among the natives. My mother, well-versed in Zulu lore, was adamant on the subject of the sex of those who cared for her children, and equally insistent that the boys leave for the United States before endocrine changes might make them susceptible to the blandishments of the adventurous native "maidens." There is no Zulu word for virginity.

Schooling was a hit-or-miss affair. My preference for the out-of-doors and an eel-like propensity for escaping confinement, left me fairly free to explore the bush from the age of eight. Bird collecting commenced with a slingshot, from which I progressed to one of the excellent British-made air rifles, later with more training to a shotgun, and finally to a 7-mm German rifle. In my mature years, I have almost entirely substituted the camera and field glasses for the rifle or shotgun, although I have no objection to hunting by those who wish to engage in the sport. Where effective predators are absent the annual surplus of wildlife should be harvested by man rather than be allowed to out-reproduce its food supply and die the inevitable lingering death from disease, malnutrition, or outright starvation. In the artificial world that man has created around him I would regard it as inhumane to substitute this fate for the quicker one of death by a predator or the weapons of man.

After leaving South Africa as a boy, I returned to the Continent twice, to enjoy natural-history studies with a more sophisticated viewpoint, although I cannot say that there is as much pleasure in these advanced studies as there was in my youth in the unplanned observation of nature or the collecting of previously unfamiliar species of animals.

Much of my understanding of the native has come from my parents, and particularly from my American mother who was also born in South Africa. My maternal grandparents were missionaries, the Reverend and Mrs. Henry Martin Bridgman who braved the six-month voyage by sailing vessel from Boston, Massachusetts, to South Africa in 1865. On arriving off Durban (also sometimes known as Port Natal) they came to anchor off the coast in the open roadstead and were then lowered into a rowboat which transported them into the bay and its settlement through the rough seas that invariably broke across the shallows and the sandbar at the harbor mouth. This precarious method of debarking was still practiced forty-five years later, although by that time larger boats, known as lighters, were often employed. As a small boy one of the thrills of landing or embarkation (from or to, by then, steam vessels) was stepping into an enormous wicker or withe basket and being yanked into the air by a winch and cargo boom, then lowered to the careening deck of the smaller vessel. The degree of shock of the final impact was determined solely by the skill of the winch master as he attempted to judge the relative and completely independent motions of the rolling vessel and the wave-tossed lighter or, later, the shallow-draft passenger tugs. The sides of the basket were so high that no one, least of all a small boy, could see what went on or when to brace for the final bump. Those were tense, exciting moments, although adults may have experienced more terror than excitement.

At the time my grandparents landed on the alluvial plain for the first time where there is now the beautiful modern city of Durban, the only accommodations were thatch-roofed, mud-walled buildings. On the slopes of the bush-covered hills back of the harbor, now the magnificent colorful residential area known as the Berea, there were still fresh elephant pits that only a short time previously had been catching these enormous and valuable animals. Monkeys, pythons, mambas, and much of the original small fauna were still plentiful in the suburbs of the city.

Following their landing, my grandparents joined a group of still earlier arrivals, and after spending some months in intensive language studies, and in assembling cattle, wagons, food and gear, they were dispatched southwesterly in search of some surviving natives in an area or community of sufficient size to warrant the establishment of a mission station. Such communities were then rare because of the massacres that had been continuously carried out by the Zulu armies only recently vanquished by the white man.

It was only after several months of trekking that my grandparents discovered some kraals in the valley of the Umzumbe River. Here, in 1865 on a ridge overlooking the river and valley, they built a mud-walled dwelling. Oblong openings the size and shape of tea trays were left in the walls as window openings; at night, "tea trays" were dropped into slots that held them in place over the openings, protection against possible marauding predators, bipedal or quadrupedal. Although the original house burned down, the homestead was occupied by my grandparents and their descendants or relatives almost continuously for more than sixty years.

At that time large game was still abundant, although the largest—elephant, rhinoceros, buffalo, hippo, and lion—had been destroyed in this area. Crocodiles were numerous, but the rivers were still free of an equally dangerous, though unspectacular little animal, the microscopic worms that cause *schistosomiasis*, a very painful renal disease that may be fatal to man. Today all crocodiles in the Umzumbe River have been destroyed (although they remain in Zululand), but the disease has spread and infests every stream and pool in Natal.

My mother and her brothers were born at Umzumbe or at nearby mission stations and they each spent much of their preadolescent lives there. It was there that the younger members of the family engaged in extensive bird collecting, my mother being the one duly delegated to do the hard and unexciting work of skinning. The collections ultimately were sold to museums and private collectors for enough to help educate her brothers for their various professions, M.D., D.D., and D.D.S., and also aided in providing a college education for my mother who thereafter became a missionary housewife. Through these useful experiences she was later able to help teach me the now old-fashioned but still useful technique in the preparation of bird skinning, but the hope that must have been in the back of her mind, financial aid for educating her own children, was doomed to disappointment because the 1880's market for avian curios had collapsed before 1900.

Despite the bird-collecting experiences, none of the family became interested in natural history beyond the first stages of collecting, preserving, and naming the animals, and only native names were available and attached to the specimens. My own early interest in the lives and activities of birds was confined largely to observation, hunting, and rather desultory collecting. But my love of the outdoors was so satisfying, that until I was nearly fourteen I had successfully resisted all parental attempts to educate me. The educational residue of those years was a love of undirected reading. The books I read at that early age were invariably classics, and I suspect that because of my resistance to formal education, my parents deliberately planted these volumes in likely places. Other than reading and a little arithmetic, my education consisted largely in the acquisition of practical, first-hand natural history acquired through the hours spent in the bush while I should have been learning more conventional things.

In uninhibited Africa, even a small boy gets many glimpses of the world beyond his comprehension, and it was not until years later that I fully appreciated some aspects of human natural history that were constantly thrust before me, such as the clandestine activities of men and women who during their beer drinks and dances or weddings slipped off into the grass. From my vantage point, usually on horseback, and with some little concealment, I watched these couples as they played together, sometimes very briefly, and then sank from sight for an interval hidden by the tall grass, to arise, separate, and go their devious ways to rejoin the dancers.

A small boy's unappreciative sense of mischief often led me to ride toward the sounds of native revelry and watch these goings on. At other times, while silently fishing along the bush-beset streams, I watched other love-making scenes in these idyllic surroundings. From observations of barnyard animals I perceived some hint of the meaning of these activities, but until later I gave them only what may be termed academic appreciation—the sight of native bulls fighting or even the warfare of the termites was usually far more satisfying to a small boy's love of action. With the onset of adolescence, when I might have gained deeper insight into these bits of native lore, my parents hurried me back to the New World, a strange world to me.

My father, who was an enthusiastic musician, upon meeting my mother in New England was side-tracked, attended theological seminary, and became a missionary. Because of his insistence that I engage in some gainful activity to earn my first firearm, at the age of nine or ten, I stumbled onto the nesting habits of the Nile monitor lizard.

As a means of earning money, my father had given me fifty young chicks to raise to fryer stage, their feed bill to be deducted from the price he would later pay me for the chickens. Fortunately, I had witnessed not only termite battles, but also the avidity with which chickens at any age glutted themselves on termites. From this observation it was an evident conclusion that free termites from nearby termitaria, known to us then as white-ant hills, might be substituted for the more costly corn-meal mash and cracked corn.

Each day thereafter wheelbarrow loads of termitaria packed with luscious termites were broken up in the chick yards. I have never seen such amazing gorging nor such prodigious growth rates, but the most lasting dividend came from the accidental discovery of lizard eggs buried deep in these ideal incubators. The realization that I had seen something that no one else had ever discovered, so far eclipsed the excitement of anything I had witnessed in human affairs, that the event undoubtedly contributed to my later professional interest in natural history rather than human biology. That this observation was wholly new was attested by the fact that I never found a Zulu, witch doctor or layman, who knew of this habit, and our reaction in the mission was that if the Zulus were ignorant, so must be the whites. Indeed, even some eighteen years later, the habit of lizards to have their eggs hatched in termitaria was still unknown and its reporting served as part of my Ph.D. requirements.

The backyard discovery of such an interesting segment of a fairly large and well-known animal's activities is still typical of the wealth of new discoveries awaiting a naturalist in South Africa. Despite the many excellent studies by South African zoölogists, much exciting work remains to be done. A free-roving naturalist with no teaching schedules or committee meetings to fragment his time, nor directed research to command his energies, can so enjoy his stay in South Africa, or any part of Africa for that matter, that one wonders why Europe and its laboratories have an apparently greater appeal for our vacationing sabbatics than Africa with its fascinating possibilities of random discoveries.

Most books on African wildlife deal with its spectacular big game and its fascinating larger birds, and only to a lesser extent have the small mammals, reptiles, and amphibians been accorded a share of nontechnical publication. The books that are concerned with these animals usually discuss taxonomic or distributional problems, or they are scholarly works that provide little of the intangible feeling of the country as a whole.

The present book is not a biological treatise. I have chiefly intended to discuss certain aspects of natural history in such a way as to invoke so far as this is possible in words, the mingling of sensory experiences and their association with the smaller elements of wildlife—the kind of thing that one finds in the farm woodlots or remaining bushlands. Even where the larger animals are discussed, I have attempted to shift attention away from those experiences that are usually emphasized in works on big-game animals, to provide an admixture of comment intended to give the feeling experienced among them, and to recreate the aura of the country in which these animals live.

The rapidity of technological changes that have taken place on the Continent in less than man's average life time has been enormous. Progress includes modern surgery, insecticides, metallurgy, chemistry, and nuclear physics, which give us so much of our excellent technology. But above all, the forces underlying the serious environmental changes have come from biological knowledge as applied in medicine and from the discovery of the new drugs. This new knowledge resulted in the bursting population growth, the ensuing increased consumption of renewable resources, and the destruction of nonrenewable goods at an appalling rate. Most of us are unaware of the speed with which man is advancing to the brink of depletion of many of the things we consider essential to our well-being. In little more than half a century the rapidly expanding consumption of petroleum has brought us to a point where we can foresee the end of large volumes of this essential commodity. It is possible that the youngest people now living may see most of the rise of petroleum technology as well as its decline almost to nonexistency. In view of the seas of oil that one generation has consumed, we have gained a picture of unending supplies, but for any nonrenewable resource the rate of our consumption is equaled by the rate of depletion.

The rising flood of humanity, and its impact on Africa as well as on the rest of the world, has led to inclusion into this book of sections dealing with humanity and humanity's hopes for the future.

As our technology advances and our population increases, time is running out for those who wish to see unspoiled nature, to describe it, or to draw biological conclusions from the amazing African scene.

1

IMPRINT OF A BARE FOOT

The spread of Western civilization in Africa is visible almost everywhere, but to those of us who have watched part of the land for even a few decades it is most obvious in the changing efficiency and speed of transportation. Impressions of the old days color the thoughts of prospective visitors so that they are prone to imagine the southern tip of this immense continent as still remote, inaccessible, and hostile to strangers. But the era of sailing ships when this was true and when the voyage from the United States often took six months, has long since passed. Nevertheless, the romanticism of those days still tints the mental pictures of those who accept the movie or television versions of Africa or who have read chiefly tales of hunting and adventure.

Thirty years ago it took fast steamers almost three weeks to make the trip. Today ships still require about the same time, but travel by air takes only two days and nights from New York to Johannesburg.

There has been loss as well as gain in the speed with which transitions can now be made from one continent and culture to another. There were rewards in the long days at sea, out of sight of land, when the ship was plunging through storms or sweating in the tropical calms: there was time for mental reorientation in preparation for new scenes. For me as a naturalist the long absence of terrestrial stimuli during the sea voyage served to

10

heighten the impact of the sensory adventures of a new world, new scenery, different odors, and the sounds of strange bird songs and of a language not heard in a long time. Man now travels within a world-wide sea of ambient air; South Africa is reached by new routes.

After experiencing flight over the center of Africa, through or high above the clouds, one realizes how vastly different history would have been had this means of transportation been possible at the start of exploration of its hinterland. The old peripheral approach by sea required a desperate struggle through the torrid, disease-ridden lowlands before the vast territories of high and healthful tropical country could be reached. It is no wonder that the beauty and enchantment of central Africa remained so long concealed.

Modern Comet jets flying at 35,000 to 40,000 feet at a speed of almost 500 miles an hour give the passenger a feeling of being gently vibrated from landing to landing and continent to continent. And even these incredibly fast planes, are only the forerunners of still faster and smoother transportation in the future.

Luxuriously, at high altitude and great speed, one passes through an ocean of shimmering air, often out of sight of the cloud-covered land. Despite the advantages, one has the disquieting feeling that air travel is too fast and too far up for comprehension of the territories that flow beneath—that flying is an inadequate preparation for the experience of entering a strange environment.

In June, 1953, the B.O.A.C. Comet jet plane arrived at Johannesburg in the late winter afternoon when the golden light of a lowering sun splashed the city that was built on gold. During the prelanding flight pattern, the plane swept in long curves over miles of impressive apartment buildings, hotels, stores, mines, and factories. The ordinarily pallid desert mountains of mine tailings glowed in the sun more richly yellow than the gold that had been extracted.

As the plane landed on the broad expanse of the airport, a large bustard fed unconcernedly nearby. A tourist-minded chamber of commerce might have arranged for some more spectacular sample of wildlife to greet a visitor, but it could scarcely have chosen a better avian representative to symbolize the local contrasts than this dignified and miniature near-replica of the ostrich. Part of the enchantment of modern Africa lies in the representations of many different ages; and this primitive native, the bustard beside the jet plane, was only a small reminder of the intricate interweaving of the old and the new. It provided a hint of the complexity of the new historical tapestry that is so rapidly being woven into the history of this ancient continent; and the speed and irregularity of the process prevent a normal evolutionary adjustment which might have produced a peaceful coexistence of blacks and whites.

As I drove from the airport to the railroad station, the city provided a vivid preview of the changes that are rapidly spreading outward from it, over the lands that were so recently called the Dark Continent. From the air there had been no warning that within the city and throughout the surrounding farms, grazing lands, and mountains, the black man was still the "prime mover," even while he was doing what he must to comply with the needs of the white man's capital and the white man's power. From the air the view had been wholly that of a modern city. An uninformed visitor could have scarcely imagined that the city and all that it symbolized for the future, had grown in only slightly more than fifty years to its present imposing stature by the sweat and muscle of the black man, the lubricant of foreign capital, and the white man's technology. In the brilliant light reflected from the ground there had been no hint of the thousands of black men swarming through the city streets and milling about in the surrounding rusty-looking slums, which from the air could scarcely be distinguished from the usual metal offal that accumulates on a city's outskirts.

For peace of mind I decided to concentrate on only the first sight of the city as seen in the reflected golden glow, and to ignore if possible the more intimate scenes that revealed the great wealth of a few and, by the standards of Western societies, the abject poverty and bottomless misery of the many.

In South Africa the line of cleavage between the two extremes is almost wholly incised on the basis of color. Extremes of economic conditions cause trouble enough, and so do racial differences; when the two are combined, as they are in this country, only a slight spark may be needed to send the white and black into mortal combat.

At the railroad station I took a ticket to Durban and farther down the coast. I soon noticed, just by looking through the window at the highland country, how under the competition of railroads, automobiles, and planes the ox wagon had practically disappeared. Even on the farms the internal-combustion engine had taken over the duties of the ox, which in not too long a time will disappear as completely as the mule and draft horse in the United States; in time, so also will pass the men who have used the animals in "treks" or on the farms.

I was to see later that at least in rural Africa the methods of travel are still a blend of the ancient and the modern and that ox wagons are still used in small numbers and for special types of work—for short-distance hauls of heavy loads on some farms and sugar plantations and, more rarely, for cross-country work in some sections where the roads are very poor or entirely wanting. There the oxen still excite the curiosity of foreign visitors who have read of the part played by this former form of transportation in the development of the people and the Continent.

Leaning back in my compartment, I recalled some personal experiences of a brief five decades ago, which may convey some vicarious understanding of a vanishing, primitive, but charming life.

To some people the word ox is now almost meaningless. Few people in the United States under the age of forty-five have ever seen oxen at work, and hardly anyone has worked with them. The work bullock probably antedated the work horse, and for long it was a contemporary of the horse and mule, but its use was limited to a few small areas of the United States and its passing has gone relatively unnoticed. Even those who have worked with the patient, plodding American draft ox can have only a partial conception of the widely used, long-horned, and more slender African draft animal. The African breed of ox characteristically exhibits as many temperaments as mules or horses, and, although it is difficult to believe, many are inclined to skittish excitability, though in a heavy, bovine way.

As I was taken effortlessly, swiftly, and comfortably to the railhead and port of Durban, my mind flashed back to the time of early travel during my childhood. Going to Durban from the missions in those days was highlighted for me as a small boy by camping along the wayside and by the still-remembered delicious flavor of the old standby picnic foods, boiled chicken, rice and gravy, plus "trimmings," or in hot weather the even more delectable chicken jelly and rice, kept cool from early chilly dawn through the hot hours of the day by many insulating wrappings. The family would travel by the only means of group conveyance available, the big wagonette drawn by sixteen active and wiry oxen. The real propulsive power, however, seemed to come from the leathery lungs and brassy throat of a Zulu driver. The daylong trip to Durban included a regular routine of predawn awakening, then rest through the noonday hours, and a late-night arrival in "town." Yet the total distance from home, Adams Mission on the Amanzimtoti River, to Durban was only seventeen miles, now normally a drive of thirty minutes.

Long before the sun would turn the eastern horizon to a lighter shade of night, we children would be roused, and our fumbling fingers would be helped with buttons and the donning of unaccustomed shoes and stockings. Overwhelming childish excitement was not slow to follow dressing, for from outside the house came the sounds of most thrilling turmoil: the cracking of the long ox whips, the shouts of the "boys" (any native male, from youth to old age) as they inspanned the lowing oxen, and the clacking of their wooden neck gear. In the dim starlight, silhouetted forms of men and cattle moved; inside the house, dotted with the flames of burning candles, sounds of scurry and bustle, laughter, and contagious excitement prevailed.

For me, I recalled, this fussing over clothes when all was confusion became an agony and was scarcely endurable. So, with the last details of dressing completed, and the restraining hand removed, I would dash out into a cold morning fresh with dew and the scents of damp earth and wet grass.

Inspanning was fascinating. There were always refractory oxen swinging their long horns threateningly at the unconcerned natives, and the crack of the whips with their *i'nfosi* or bushbuck-skin tips, as more oxen were hurried to the waiting yokes and chains. But best of all was the chance to chatter with the little "leader" boy shivering in his scanty *mutsha* (skin apron) and threadbare undershirt. He was right from the bush, *he* knew how to set traps which would actually catch birds, and his snares were equaled in deadliness only by his skill with the rubber slingshot. He was a real person doing important work, though of course he was not as important as the big men stepping boldly in front of the unwilling oxen, the men who, by strength of muscle and lung, could persuade one more ox to move forward one more inch until the neck strop at last could be fitted.

To me, as a small boy, these men were real heroes, and long before I had seen a policeman or streetcar conductor, I had resolved to grow up to be a bullock driver or at least a *skulufa* boy, the brake tender who ran behind the wagon and wound up the screw-threaded brakes until the rear wheels slid and the wagon skidded perilously. This was certainly fitting indoctrination for a sympathetic understanding of these men in later years. Of course there was hero worship, the objects venerated being the fine big black Zulu men, straight, upstanding, powerful, the sons of men who had been valiant fighters. To me, however, it was their ability in handling the long-horned oxen and the calm, steady carrying on with their work that excited my admiration.

Those early dawns were often frosty, and so full of shivering excitement that when the call to breakfast came even cake seemed uninteresting. At last, the family would rise from the table, and, after blowing out the candles, which scented the room with the odor of smoking paraffin wax, we would move out to the waiting wagon.

The first long rush from the mission house down the steep road toward the level plain below was a fitting climax to the predawn activity. As we dipped downward to the valley, we would look off to the dimly glowing east and see, in the river bed below, faint veils of mist clinging to the colder spots above the tepid water. Sometimes in the light from the swinging lanterns the grass in the valley sparkled with bluish gray, and we would know there had been frost that night. The Zulus would laughingly shout *"umqoqwane"* (frost) and trot unconcernedly on, their bare feet dusting powdery ice from the long grass.

At the river would come the usual battle with the old Volo, one of the wheel oxen of the first span. He invariably refused to cross that river, and when the "boys" met the refusal with the customary prompt treatment—a sharp bending of his tail and a hard bite on the painfully tense bend—he frequently countered by lying down in the water.

By the time the long span of oxen was pulling over the rise beyond the river, daylight would appear, changing the weird shapes of trees back to familiar outlines. With the light would come the odor of fires burning in native huts: wood smoke is still, after fifty intervening years, the fragrance most evocative of memories for me.

As I rode along in the train, no memories were quite so vivid as those of the long wagon rides through the cool of the day and, later on, the smell of dust and the gasps of hot, tired oxen panting against the yoke, the sound of the iron-shod wheels on sand, a soft, steady grind, or, on hard gravel, crunching and sputtering. The oxen would be urged forward by the guttural shouts of the Zulu driver and blows from the doubled whiplash. The *voorlooper,* or leader boy, would pull on the looped rawhide thong attached to the first pair of animals, or race along the lines of straining beasts, applying a needed stimulus where the angle of a yoke showed evidence of shirking. When a long day was dragging to a close, and the sinking sun lighted the hilltops, throwing bush, trees, and valleys into deep relief, the evening outspan would come as the tired oxen plodded to the top of a last hill before reaching camp.

Occasionally there were intervals of real danger, times to be long remembered. These adventures came suddenly, with no warnings. Once we came to a tongue of "bush" cutting across the road which had been tunneled through the dense thickets and vines when the oxen, which had moved lazily forward under the draped branches, suddenly stirred uneasily as they sniffed the air. Fatigue and training forgotten, they leaned into their yokes and, bellowing savagely, surged forward into the open. There was no stopping that wild stampede. They fled forward, their eyes bulging with fear, horns clacking and tails lifted, each plunging stern flaunting a banner of fright. Behind the sixteen yoked and supposedly placid cattle, a heavy, tented wagonette lurched, skidded, and bumped, while three gleefully shrieking children and a white-faced terrified mother clung to the swaying sides.

The guiding *voorlooper* was nowhere to be seen, the brake boy had been left behind, but the driver had dropped his long whip to race madly along behind the wagon, spinning the brake handle around and around until the rear wheels stopped turning and slid, smoking, along the ground. Bumping over the rough road, the wagonette tobogganed through sand, over rocks, and grassy hummocks, jolting wildly and threatening to capsize. The oxen swung from the road, and as the wheels reached the deep, trampled grass, they glided swiftly over the slippery surface with the oxen gathering speed.

There was nothing to stop the terrifying rush, and the stampede headed toward a deep ravine. Within twenty feet of the ravine's edge, a wheel ox stumbled, fell, and, half turning, snapped its yoke. Momentarily he had attained freedom, but the next instant the careening front wagon wheels were on him. Horns and yoke tangled in the flying spokes and somehow slewed the wagon away from danger. With three of the wheels locked or obstructed, the heavy wagon slowed the oxen to a stop, and they stood with heaving sides, uttering low, moaning bellows as they awaited their drivers.

From beginning to end of the incident, not more than five minutes had elapsed, yet the so-called lazy oxen had broken from a shuffling plod to mad speed. They had pulled a loaded wagon, with its wheels braked to a standstill, over a rough road, open fields, and along the brink of a ravine. They had been brought to a halt by mere accident, a deep ant-bear hole in the path of a wheel ox. The cause of the stampede was no less a matter of chance. In a country where leopards had not been seen for many years, one had suddenly passed across the road not far ahead of the oxen.

I was brought back to reality when the train approached Durban. I could see that the city had spread, especially its slums and factories, and also the sugar cane enveloping many hillsides in a uniform mantle of green. Birds appeared to be fewer, but the flocks of little cattle egrets seemed to have increased in size and number, and many could be seen stalking along beside occasional cattle or riding on their backs like miniature white ghosts.

In Durban I changed cars for the south-coast railroad that would take me to Umzumbe. After we left the metropolis we passed what were once great mud flats, rimmed and ornamented by mangroves, with here and there on higher ground the early plantings of sugar cane. The leisurely, crawling pace of the train was like a lingering relic of the days of wagon treks. Much had changed on the south-coast railroad from Durban to Port Shepstone, but doubtless many changes were too gradual to be noted by permanent residents. But I could see that the rate of change must have speeded up and increased the pressure on remaining bush and fields, rivers, and native wildlife.

Still, as the train puffed laboriously up a steep grade, where the Indian Ocean appears off to the left, and slowed to a crawl on the curve near the summit at Umbogontwini exactly as it did years ago, I could examine the dusty banks of the red-soil cutting and in the fine dust see tracks of the resident wildlife. But even this will not be for long. Progress is moving fast, and even this little corrugated-iron lean-to station will soon be "modernized" and overcrowded like the land.

There were man's tracks too. A newcomer, like Robinson Crusoe, might have been startled to find himself gazing at the clear-cut print of a large human foot. That simple incident was symbolic in its way of the beginning of Africa, a land that in spite of its cities, and shoes, is still the country of bare feet, where footprints in the dust of the road or in the hardened mud near a water hole take on new meaning.

There was really nothing remarkable about the presence of footprints, of course, and bare feet should have been natural. Nevertheless, the sight of a bare footprint near the railroad track characterized the still primitive social development of this land and its people.

This introduction into Africa, the recognition of the imprint of a bare foot, can be very significant to a newcomer. The traveler in remote areas can see these tracks wherever he goes. He soon comes to notice a shod footprint with more surprise than he would the natural imprint.

Since the train had left metropolitan civilization and the world of the white man, I expected to see wild monkeys walking across the tracks. For decades at Umbogontwini, only a few miles south of Durban, black-faced, sad-eyed monkeys have wandered aimlessly across the tracks, from the dense bush on one side to the gardens growing between the railroad and a factory on the other. It was always here, even as far back as the days of the ox cart, that monkeys could be seen. As the decades passed, so did the ox wagons and the horse and buggy, but the monkeys remained. The trains will continue to run, but the days of these monkeys are limited, and within a few years they probably will have disappeared from this rapidly developing section. Only the presence of what was once an explosives factory and its need for surrounding space have saved them for several more years. In city parks and wildlife sanctuaries they will live on in semidomestication, but elsewhere they are doomed to go.

Even as in the old days, the train south from Durban barely managed to struggle over the crest of the Umbogontwini grade. As it slowed I looked for these friends of years past. I was rewarded, for at this old familiar spot, a black-faced, grayish monkey sat leaning back against the bright red lateritic soil of the cutting, chewing on the contents of his cheek pouches, with legs crossed, hands back to brace himself, calmly contemplating the slowly creeping train.

As I scrutinized every blossoming kaffirboom (coral tree; *Erythrina*, native *umsinsi*) in passing, a movement among the branches revealed an entire band of a dozen or more monkeys: a big dominant male, several females and lesser males, and the little tots. They were feeding on the blossoms and nectar, as well as on unwary insects; they reached out their little black hands, pulled in cluster after cluster of the crimson blossoms, and buried their faces among the flowers. The vivid crimson, the black hands and faces, the gray bodies combined superbly with the gray of the coral-tree trunk outlined against a clear blue sky.

Out in the wild country one does not often nowadays see monkeys strolling casually in the open or even poised in the shelter of their tree refuges. If seen at all, the fully wild monkeys, accustomed to constant persecution, are more likely to disappear in a flash of gray among wildly thrashing foliage, followed by utter stillness. They can disappear more completely in even limited cover than other animals half their size.

I recalled one mid-October morning in the 'twenties after a night of rain and cold. The day was still overcast and chilly. The dew or possibly undried rain soaked everything that moved through bush or grass, and through the thin cloud layers above there came little warmth. In a sunny clearing in the bush I saw a troop of monkeys sitting along the railroad tracks, with only a couple of the group still prowling for food in the nearby grass. The rest of the troop were sitting quietly on their haunches with their hands on the iron rail, or sitting parallel to it, with both hands, feet, or buttocks pressed to the metal. Thinking that there must be a reason for such uniformity of behavior in the troop, especially for their attachment to the tracks, it occurred to me that the temperature of the ground and rail were different—it was no surprise to discover that the iron was almost too warm for comfort while the earth was still damp, clammy, and cold.

For most of the south-coast line between Durban and Umzumbe the railroad followed a tortuous course along the seashore, in and out between bright green fields of grass or even greener sugar cane. For miles, except where broken by the blue lagoons at the river mouths, the dark green bush of low trees and clinging vines, *isigudi* or wild "bananas" (*Strelitzia*), flaming aloes, and graceful palms hemmed in the right of way along a shore of tawny sand and black rocks, bordered by the indigo ocean.

The crowds of Zulus, packed into the third-class compartments forward on the train, represented the lowest economic stratum among the travelers. They were ignorant, exploited, and oppressed. By many they were also regarded to be "sinful" or "degraded," but nonetheless theirs was the only part of the train from which came sounds of happiness—singing, the strumming of a guitar, the jouncing rhythm of a mouth organ. The sound of chatter was mingled with hilarious laughter, whereas farther back in the better accommodations, the native chiefs and teachers (the better-dressed and undoubtedly wealthier travelers) were solemn and reserved. Still farther back in the first and second-class compartments reserved for the whites, there was, if possible, even less gaiety except among the children; and repression, boredom, and stuffiness characterized the faces.

Seemingly as a man acquired position that was held only precariously by a narrow range of comportment and money, spontaneous happiness expressed outwardly and shared with others diminished.

If we whites could acquire the gaiety of the native merely by losing our possessions and forgetting the values we place on social status, the exchange might be worthwhile. For us it is always surprising to watch even the poorest of the natives and see their irrepressible pleasure in life. Even when actually hungry, and in the cold gray light of early day, they are prone to sing and dance, or indulge in practical jokes. At dawn, after an all-night ride on the hard wooden seats of their train, they will wake and poke their heads from the windows to sing, shout, or exchange ribald jokes with any native woman in sight.

During the morning I heard an unusual crescendo of shouting and whistling from the native coaches. I looked out and saw the unforgettable sight of the stark-naked golden bodies of two nubile girls, glistening with drops of water freshly splashed on their breasts and shoulders and framed under a graceful palm outlined against the blue of the Indian Ocean. In their solitude they were daring convention by waving their hands and exposing their bodies, and with perfect safety to themselves causing an erotic storm among the native men massed in the speeding train.

Once more, as we approached our destination, and I visualized the last leg of the trip from the station to the inland mission, my thoughts returned to early days. When we would leave the train for an inland trip and the panting oxen or sweating horses reached the last hilltop leading from the coastal bush to upland home, we would always cast a backward glance at the blue ocean, silver-flecked and purple-shadowed, stretching out and away until on the curve of the horizon the indigo sea and turquoise sky blended to lavender distances. With this last glance we relinquished the familiar things—trains, telephones, hotels, and convenient stores, but it was only by leaving the America-like towns and cities of the hybridized margins of the continent that we entered Africa proper.

It was this Africa that I had come to see. It was the Africa that started after a few miles through a land where all things seem almost made in miniature—trees, hills, and dwellings—when suddenly after a turn of the road the sea and shore are lost to view, the free-flowing breeze is shut off, and Africa emerges, Africa of the senses, of beauty and ugliness, and charm and repulsion; Africa the land of the blue sky with clumps of fleecy white clouds; Africa which throughout history has engulfed its intruders. Only along the lower Nile or north of the Sahara has civilization maintained tenuous control. Elsewhere, Africa has resisted encroachment, and with the slow but effective weapons of time and human fecundity has blended the temporary invader into oblivion.

These were my thoughts as I left the train at the Umzumbe railroad station to revisit the mission founded by my grandparents—the site of my childhood.

As I drove up to the familiar grounds of the mission station, from among the flamboyant scarlet blossoms of the umsinsi coral tree shading the entrance door of my old home came the well-remembered sounds of a Cape ring dove.

The mission was to be my base of operations for trips to various areas, especially the game reserves, and would forever be the focal point of my thoughts concerning past, present, and future of the African scene.

2

HLUHLUWE AS IT WAS

HUNTING IN THE RESERVE

Nowadays good roads and the automobile make it possible to reach the Hluhluwe game reserve in a few hours' drive north from Durban, but thirty years ago it was necessary to take a train, old-fashioned and primitive, from the city to Somkele, Zululand, which, at that time, was the end of the rails. The 150-mile trip would take all night, and the train was supposed to reach Somkele by 10 A.M. It rarely arrived on time, but a few hours' delay was not important because the necessary ox wagons, true to custom, could be counted on to arrive even later. From Somkele it was a two-day trip by ox wagon to the Hluhluwe valley, and one night was usually spent camped along the roadside or at a mission station en route. In many respects I learned so much about animal life and the country in those days that an account of the old hunting ways might profitably precede the chapter on game photography today and thus give me a chance to indulge in reminiscing.

The approach to our usual destination, the area just outside the Hluhluwe reserve, passed for many miles through uninteresting territory which even then had been cleared of large game animals. But closer to the sanctuary the terrain became

23

progressively rougher and more scenic until eventually we would reach the steep-sided hills enclosing the valley that constituted the principal part of the reserve. On a high ridge from which we could look down into the bush-streaked valley we always changed from oxen to donkeys to avoid loss of cattle through the tsetse fly and nagana, the bovine sleeping sickness. While we switched from one form of transport animal to the other, we could see other ridges and high hills ranging far beyond into the bluish wintry haze, and in the gaps between the ridges were the distant landmarks of the wildlife reserve. Summer trips were avoided because of the prevalence of malaria fever.

In those days, we usually camped close to a stream shaded by a few African sandalwoods, the *umtomboti* trees, exceedingly fragrant euphorbiaceous hardwoods, incense-like when used for campfires but reputedly dangerously poisonous if used for cooking fuel.

Invariably camp sites were chosen not far from a native family village, a kraal, where we could get guides, gunboys, and carriers in exchange for a share of meat. Today only money procures their services.

There was always the busy work of camp routine, and when the tent was finally pitched and the boxes stacked, night would have fallen. Then as we sat about the campfire for an hour or two, we talked to the native men and learned something of their philosophy and of their wildlife lore.

Because it was necessary to start before dawn, we usually turned in early under heavy blankets. In the clear air of a South African winter's night the day's accumulation of warmth was rapidly radiated into space, so that temperatures dropped sharply after nightfall and often resulted in local radiation frosts.

The world outside the tent always was wide awake. A short distance from camp a few Woodford owls, *mabengwane*, would call and answer, while close at hand, under the edge of the tent, a cheerful cricket always seemed to chirp a welcome. Listening to the cricket, enjoying his vivacity and noisy persistence, was pleasingly relaxing, and the world of reality would quickly fade into sleep.

During the night, a few hours later, when the full moon was setting, its pale light would filter dimly through the leaves of stunted flat-topped thorn trees in front of the tent. The cold and watery light accentuated the chilly frostiness of the night wind, and even in the warm blankets it was often difficult to close all the numerous openings down which cool trickles of air seemed persistently to find their way. If one was disturbed by the cold, and awake, there would be nothing to do but listen to the sounds of night.

Down the valley the sharp whistle of a reedbuck could sometimes be heard, the sound ridiculously like the sharp squeak of a squeezed rubber doll. Eddies of air may have carried some scent of the camp to him, or possibly a rival male may have threatened invasion of his private domain. Up on a ridge, back of camp, a zebra might bark, making a shrill, querulous sound like the yelp of a sea lion, mellowed by distance to clear, musical tones lending enchantment to the African chorus.

Night was always a time of fascination, and sounds from the throngs of game came through the quiet air with greater clarity than during the day. Often the distant, melodious baying of a hyena roused us from sleep, and for several minutes the sounds would come regularly, growing fainter with the distance until the silence settled down again, heavy and mysterious. Although there was no actual danger, it was at such times that one most fully appreciated the basic survival value afforded by fire and felt glad that man had learned to make homes and keep fires burning through the night.

For one attuned to catch every sound, the night became full of mystery; vague rustlings heard here and there close to camp, or the sounds of a heavy body slowly and cautiously moving through the bush twenty yards away, and again faint snorts in the distance—or long, tense moments of absolute stillness.

When at length the illuminated dial of a wrist watch showed 4:15 A.M., it was time to be stirring if this day of hunting was to be a success. But it was very cold, and the blankets held an irresistible appeal. The cook was called, and from the boys' quarters not far away a sleepy voice would mumble, *"Kona manje, n'kosana"* (Right away), and knowing that the cook had been aroused, instructions would be shouted to have the fire built up, water heated, and hot coffee made to start the new day. These were camp luxuries that gave added pleasure to "roughing it" in Africa.

While the world was still dark, in the half hour between the setting of the moon and the first glow of dawn, we ate breakfast in sleepy discontent, crouched over the fire for warmth.

We would spend the day among the hills at the edge of the game reserve. This boundary hunting, though seemingly unsporting, was actually essential in order to keep the game within the reserve and out of competition and—even more important because of the danger of nagana sickness—out of contact with cattle on the surrounding farm lands. Spread of nagana through wandering game threatened—and still threatens—the safety of the reserves themselves.

We had to be in a position to intercept the game on their dawn return to the reserve up-wind. Here, as in any other locality near a refuge, the animals had learned to move out of the sanctuary chiefly at night, and then toward morning and during the first hours of daylight, to graze back toward safety. The more timid animals would reach the sanctuary before daylight, and the bolder and more restless individuals that had strayed farthest in search of fodder returned later, or even loitered in the shade of the bush near the borders.

When the animals just outside the reserve became aware of danger, they invariably dashed toward safety within their sanctuary. With the echoes of the first shot of the day, long lines of wildebeest, zebra, and other game would trot single-file into their haven.

During the northerly wind, *n'yagato,* which blew from the reserve toward the area chosen for our collecting, many herds of game would not stir from the reserve at all but would work up-wind, their favorite maneuver for safety, and then graze deeper and deeper into the reserve, even though the food there was sparse. This fact became increasingly more evident in that area, and the only notably good days were those when there had been a southerly wind, *u'nigizim* ("bring sheep," i.e., clouds). When this wind had been blowing into the sanctuary all night, we could count on seeing herds of all kinds of game grazing or working back to their refuge in the morning. Obviously at night, probably in the daytime too, the animals were prone to graze up-wind in order to scent any lurking danger. In this way they were safe from molestation in front, and they used their eyes to detect danger which might approach them from against the wind.

The habits of animals are inevitably changed by life in a game reserve, and by the safety that experience teaches them lies within. In most other respects, however, the game is probably as it always has been, and reacts in much the same fashion. It is surprising that animals "know" so precisely where safety lies. Time and again they dash determinedly past a hunter, even at close range, but only when in terror of visible and obvious danger. Apparently they are willing to take a chance in order to get back to the reserve. Sometimes they rush onward for a hundred yards after passing the hunter, but unless hard-pressed, they slow to a trot or lope as they approach the boundary, and stop when only about a hundred yards within the invisible lines.

Wildebeest and Bushbuck

Sometimes, during our hunting trips in the 'twenties, herds of wildebeest and less often zebra that had gained a strategic position near the border, would trot a short distance, turn and eye us, trot closer to the reserve, look again, and by slow degrees work back into safety. If at any time during this progress, we made an attempt to get ahead and intercept them, they would race to maintain their relative position. It was almost impossible, even with careful stalking and ideal conditions, to head them off from their refuge once they had seen a hunter.

I vividly recall one specific wildebeest and bushbuck hunt in those days. By making an early start I planned to be with my Zulu boys on the crest of Mahlungulu hill by dawn, and follow up-wind along the ridges paralleling the Hluhluwe reserve boundary, marked only by white, diamond-shaped beacons and the line-of-sight between. Leaving camp in the care of a boy, who huddled over a fire and in the dim light looked like a huge, distorted silhouette of a toad nursing an orange flame, we set out for the hills.

It was still almost night. Over the mountain to our right the sky seemed to be somewhat paler with the first touch of dawn, but all about us was the full darkness of a starlit night. We followed a game trail which headed for a notch in the hills below Mahlungulu crest. We were alert and straining our eyes for sight of game and for the dim line of the winding trail. No one spoke; there was only the steady, rhythmic breathing of the men behind me, the soft shuffle of bare feet, and the swish of parted brush to tell me they were following. Over the first ridge we went, and down into the valley where some large birds rushed up from under foot, making a loud whir that sounded thunderous in the tense silence and started hearts beating faster in anticipation of the rush of a larger beast.

The cold air had settled into the valley during the night, and in the faint starlight the grass gleamed silver, palest at the upper edges where dew and cold had not yet combined to turn this ebony bowl to the silver of frost. Our path led through this bowl, through head-high grass, and as we walked, the needles of frost sifted down on our warm clothing and turned to drenching dew. My guide was reasonably well dressed, but all the others were only sketchily clothed, wearing cotton shirts or only the traditional *mutshas* (skin aprons) of the old-time Zulu. Even with warm clothing there was a distinct chill, but the Zulus only clicked out their word for frost, *"umqoqwane,"* and stoically trudged on.

As we ascended the gradual slope of Mahlungulu, the eastern sky faintly began to turn rose, and nearby trees stood out as grotesque black shadows. Here it seemed less cold, and as we walked upward we passed through belts of warmer air, fragrant with the clean, pungent smell of earth and grass. We stopped for a breathing spell. Far away some big ground hornbills boomed their calls and answers, the first bird notes of the day, monotonous, insistent, unmusical, yet strangely appealing, like so many of the charms of South Africa.

On the summit of Mahlungulu a predawn hint of temperature changes stirred a breeze that blew steadily and cool. Since it was too dark to see the game, we could only sit and wait for daylight and while sitting, watch and listen to the magic of the coming day.

Far down below, wildebeest were grunting and coughing, either in fight or frolic. These were the only sounds, at first, but as the light grew stronger, the smaller birds began their cheerful notes. First one faint song and then another sounded, until all the bush was awakening.

The day grew lighter. As the red in the east grew more intense with the coming sunrise, the distant hills turned pink, the nearer ridges rose, and between them lay the black bush, mysterious and exciting. Far below, along the banks of the Hluhluwe River, spread the thorn bush, silent and black. Here and there were spots of bluish gray—almost silver—dead thorn trees still standing, although the bark had fallen from the trunk and twigs. Where grass fires had swept through the bush, the trees showed as black silhouettes against the pale gray ash.

As soon as the light was bright enough to allow hunting, I left my extra coat with one of the boys and started out along the ridge. Game was abundant, especially the mountain reedbuck or rooi-reebok, the *n'xala* of the Zulus. This is a beautiful little antelope, weighing scarcely seventy pounds, and resembling the true reedbuck, although it is smaller. It frequently gives exhibitions of marvelous jumping, and bounds away downhill at full speed, apparently careless of rocks and trees that everywhere form traps for unwary feet. However, at this time we were not out after the smaller animals but after the wildebeest and zebra, if these could be sighted close enough to camp for easy transportation of the hides.

The local Zulus did not mind carrying a heavy hide for miles, if they could be sure of having a generous share of the meat, but they would not eat zebra. Unlike the people of many other tribes, they abhor it, a fact that made it difficult to get zebra hides to camp.

Even after many years of trying to visualize big game in the field, the first sight of it is always fresh and new. Many are disappointed, because experience in the field differs so widely from preconceived notions. One expects to see a herd of zebras stampeding out of thick bush and looking as they do in pictures, but what usually is seen as one tops a ridge and looks over the countryside is a patch of scattered shining white dots standing in the tall grass. That they are rather large grazing animals is easy to see, but at a distance these spots look so much like domestic cattle that until a Zulu quietly says *"amadube"* (zebra), the inexperienced may let them pass unnoticed. Because zebra and wildebeest are equally numerous and have a strong affinity for each other, probably one will also hear the Zulu mention *"n'konkoni"* (wildebeest; brindled gnu) and find the attendant pointing to some dark spots scattered either near or among the zebra. At a distance the zebra looks almost pure white, the wildebeest black, but either one could readily pass for domesticated livestock.

After walking along the hill for half a mile, we dipped into a hollow, climbed cautiously up the slope and looked over to reconnoiter. At first we saw nothing, then suddenly a movement close at hand revealed a herd of ten wildebeest, none of them more than fifty yards away. These wildebeest, with their graceful recurving horns, are of all the hooved animals the most grotesque. Their long, Roman-nosed profiles make them appear solemn, but they vie with the most light-headed calves in their strange and ludicrous prancings. They have heads that seem too large for their small bodies, and the horns are puny replicas of the African buffalo's. But here the resemblance ends, for the neck and shoulders have a mane like that of a horse. Their feet are antelope-like, but the hindquarters are like those of a horse. This absurd combination is topped off by a tail almost exactly like that of a mule. The Zulus avow that the wildebeest are the progeny of a buffalo and a zebra. To support their belief, they point out the darker stripes on the paler background, and still more emphatically announce that only a horse or a member of the horse family has two nipples on the udder, and that, obviously forgetting some exceptions, all other animals have four or more. Here is an animal with many horselike traits, especially this last important feature—how, then, can the *n'konkoni* be anything but a hybrid?

We shot two wildebeest. Then I sent a messenger to the nearest kraal for skinners and carriers. By 10 A.M. he returned with a half dozen of them, followed by fifteen or twenty of their wives. The women gathered around close to the animals and helped carry on the work of skinning and cutting. In their efforts to salvage particularly choice delicacies, such as tripe, heart, liver, or intestines, they worked especially hard, for all these tidbits would go to their homes, and woe to them if their husbands decided that they had not received their share.

Friendly squabbling over pieces of meat was almost incessant, and there was continual chaffing, tale telling, and gossip. The women were dressed in the usual leather kilt and draped blankets; their hair was done up in grease and red clay; but the buzz of conversation sounded like women's talk in any country. White men who understand Zulu are so rare that at such times as these the women would forget the possibility of being understood and talked freely in front of a linguistic eavesdropper.

The conversation was loud, and the discussions drifted from the quality of some neighbor's beer to complications at a recent birth; the quantity of kaffir-corn from a certain patch of ground to the latest death; from courtship scandals to the increase of ticks; and so on through all the flirtations and seductions and other delicious topics that active and uninhibited minds could conjure up for discussion. Matters that would be whispered in most countries were discussed in loud and carefree voices.

The Zulus live close to all phases of life from babyhood to the grave. The little herder boys, from the time they are able to toddle, see all the intimacies of life, and the mating of the cattle and other animals. All external biological functions are as familiar to them at nine or ten as the rules of baseball are to the American boy of the same age. The result is that these things seem to them as natural as breathing, and become topics for everyday discussion. Sex is a diversion, frankly fun; the only wrong is in being caught by a jealous spouse or lover or becoming pregnant if unmarried.

While some of the men were skinning the animals, others were cutting down branches from nearby trees and making a mat of clean fresh twigs and leaves beside each animal. The hide of each wildebeest was carefully peeled back from the body on one side, loosened clear to the back bone and a little beyond. This side was then pulled out taut and the animal rolled back on it, exposing the other side for work. Here again the skin was treated in the same way, so that in a short time the animal was lying, completely skinned and perfectly clean, on the middle of the hide. For skinning, several men took spears, and they were marvelously dexterous in the use of a weapon that would seem to be an awkward tool. Other men had pocket knives, which they used skillfully, and all of them brought keen, two-foot-long, bush knives, which were used for chopping through some of the heavier joints.

Little time was needed for the actual skinning, but when it was completed and the cutting and drawing of the animals started, there were frequent stops for rest and food. As soon as possible, certain members of the party dropped out and cooked meat, some of which went to those still working.

The meat looked like beef in texture and color, and the natives considered it good eating. So I anticipated a good meal, but I made a mistake by failing to cut away the fat and suet. The sight of a nice fat piece of steak roasting over the coals was appetizing, but with the first mouthful I realized that wildebeest fat at the temperature of my mouth is the hardest, gluey tallow imaginable. It clung to my fingers, to the roof of my mouth, tongue, and teeth, gummy and sticky. That alone was enough to spoil any possible enjoyment of freshly roasted meat. The boys seemed to find it excellent, but they probably enjoyed my look of surprise fully as much. I could not be sure of this, however, for Zulus were always respectful, and for them to have laughed or shown amusement on this occasion would have meant rank disrespect according to their etiquette.

The Zulu normally has little protein in his diet, and he is almost always ravenously hungry for meat. As usual at the skinning, little fires were going on all sides almost before work had really started. Sometimes, when an animal had bled internally and the body cavity was full of warm, slightly coagulated blood, the natives dipped their hands in, cup-fashion, scooped up handfuls of the blood and drank the hot liquid with evident relish.

It was always a pleasure to watch the "savage" at his work —deft, coöperative, happy, and forever joking. Frequently during the proceedings some man started a deep, humming song. In the chorus all the big, stalwart Zulus joined with enthusiasm, singing splendid bass or clear tenor, each taking his part as naturally and confidently as he walked or talked. There is no race that sings more naturally than the Zulu.

As soon as the skinning and cutting up was done, the men distributed the loads of meat. I had presented definite sections to the individual kraals as compensation for the work. When all loads were distributed, the men gallantly helped to load up the women. Then, carrying only their spears and knives, the men set out ahead eagerly striding toward the beer pots at home, while the heavily laden women toiled behind.

The Zulus have never been ashamed to work and have always believed thoroughly in labor, but they believe in its proper distribution. They enthusiastically agree that the proper work for men is tending the cattle, giving instructions to the herder boys, hunting, fighting, and building homes. When this is done, they return to their hut or the nearest shade and, with a clear conscience, order their wives to bring beer so that they can comfortably discuss plans for the next project. The woman's work is finished when she has cultivated the fields or harvested, carried in the firewood and water for the day, swept out the yard and hut, tended to the cooking and the babies, worked on the skins used for clothing, and run minor errands for her husband. As in almost all primitive countries, where the men must be constantly prepared for attack against themselves and their women, the women must carry the burdens—originally probably a life-saving custom that enabled the men to spring to their own and their dependents' defense uncluttered and unwearied or distracted by the other duties. Thus, even today, women must carry the burdens, the men their arms, and the men walk ahead. For obvious reasons this custom will be slow in dying.

When the carriers had started for camp with their loads, I felt free to hunt for a needed supply of less cloying fresh meat for immediate use. With a few exceptions the larger animals are not so good for fresh meat as some of the smaller ones, and wildebeest are almost useless for this purpose. As biltong (dried venison), however, all are excellent.

I took two boys to carry any meat I might obtain, and headed down into the valley, where we could count on bushbuck or reedbuck. Hunting in most places was simple; the bush was open, often just a thicket of scattered thorn trees, and the grass was short. On the floor of the valley, however, were stretches of high rhinoceros cover. Nevertheless, this was a good section in which to find bushbuck or reedbuck and it was the shortest way back to camp.

I expected to find game abundant, but by the time we started, after the delay incident to skinning the wildebeest, it had become rather hot. All smaller animals seemed to have taken refuge in the thick bush immediately along the streams. Tracks of bushbuck (n'konka), reedbuck (n'xala), and duiker (m'punzi) were common, and here and there the spoor of wildebeest and zebra showed that those animals were headed for the reserve, probably because they had been disturbed by the early shooting.

At one place where our trail cut across an open patch of ground in the bush, fresh rhinoceros tracks showed plainly in the sand. The animal must have passed only a few minutes before we arrived, and probably an eddy of the erratic breeze had warned it of our approach. It was just as well that it had moved on, for the country at this spot was covered with thick brush, shoulder-high grass, and twisted, flat-topped thorn trees. It was not a comfortable place in which to disturb a rhino while carrying only the light rifles we were using.

In country such as this, one could be on top of a rhino before seeing him, and if suddenly frightened, he might charge.
If warned ahead of time most rhinos would probably move off
and not cause trouble. The consensus is, however, that one can
never tell what they will do. From a few observations it seems
as though the *sudden* sight or strong scent of man close at hand
is more likely to start a charge than a retreat, although a "charge"
seems to be essentially a retreat that is unfortunately aimed in
the wrong direction. Rhinos are obviously slow-witted, and sometimes one of them can be seen "doing his best to make a deci-
sion," turning, hesitating, sniffing for more information, getting
ready to run, then apparently forgetting which way he decided
to go. The presence of rhinos in the bush where one is working
adds a good deal of zest to what would ordinarily be just plain
collecting.

While walking through the more open country, I scared up
and shot a young bushbuck doe and, because I needed the meat
and the skin, I took her along to camp.

PROTECTIVE COLORATION

The male bushbuck, though smaller than our Virginia deer, is
a fighter and probably kills more hunting dogs and injures more
men than any other game in Africa, with the exception of the
carnivores and the largest of the big game, such as elephant,
rhinoceros, and buffalo. Until wounded or cornered the male is
extremely adroit in escaping detection. The horns, somewhat
twisted along their axis, reach a length of eighteen inches and
are formidable weapons, and the sharp, hard hoofs are also useful in defense. When almost all herbivores in a patch of bush
have left, it is a fair assumption that a bushbuck ram will still
be hiding there. His fondness for lying quietly and refusing to
move, even for dogs, until they are almost on top of him accounts
for the frequent killing of dogs by the bushbuck, but it is possible that because a male often refuses to run, the dogs frequently
pass by with no more attention than they would give a cow or
other domestic animal that will not run or show alarm. Only a

well-trained dog will do good work when hunting bushbuck rams, and his life is likely to be short if he pushes matters too far—he will be impaled on their sharp horns.

These comments on bushbuck apply more particularly to those living in frequently hunted areas. The same behavior may be true elsewhere, but it is not possible to vouch for it in Zululand where bushbucks rarely encounter domestic dogs but in the past often faced wild hunting dogs, an entirely different kind of animal.

Tales of the determined and courageous nature of the bushbuck antelope are numerous, and possibly as exaggerated or misinterpreted as many stories concerning the mamba (tree cobra). I am especially inclined to this belief since the terms "savage" or "vengeful" are so frequently used in these accounts. Nevertheless despite skepticism as to their "savagery," I think there can be no doubt that they differ from most of the lesser ungulates in their reaction to wounds or danger.

In my own limited experience only one of these animals, a cruelly wounded male, seemed actually to attack or "charge." Whether he would have carried this attack to its final limits or at the last moment veered off or retreated, I did not discover because at the moment it seemed unwise to delay action long enough to discern his real intentions. When he charged, it was across an open clearing surrounded by tangled, impenetrable vines, and there were no handy trees to climb, therefore self-preservation called for an immediate kill.

On the basis of two stories told by Zulus who showed me the exact place where the events had taken place, I am inclined to agree that the animals, wounded or not, can be aggressive under special circumstances. In one instance a Zulu was engaged in driving a bushbuck ram through and out of a "donga" or small ravine where the cover was exceedingly thick while a friend of his took a stand at the edge of a central ravine through which the buck was expected to appear from a game trail. Unfortunately for the friend, the ram burst into the clearing behind him, and before he could be warned the buck had charged his rear end, driving a horn deeply into each "cheek" with such impact as to carry both man and buck over the brink of the ridge. While in the air, the man drove his spear downward and back with such good effect as to kill the antelope, but he was loath to sit down for many days thereafter, seemingly the funniest part of the story judging by the laughter that accompanied the recital of these events.

In the other instance a wounded ram dashed into tall grass toward some dense bush that would have provided ample cover of the kind normally adopted by these animals. Expecting the ram to continue on into the bush, a natural conclusion because of these animals' preferences for cover, the Zulu dashed heedlessly into the grass, whereupon the ram rose and drove at him, thrusting one horn between his ribs and arm, the other into his heart killing him.

A few years ago near the town of Umzinto a ram bushbuck, pursued by dogs, broke into the open near some school children and attacked and killed one of them. Again, only a few miles away on a road passing through dense sugar cane, another bushbuck, apparently unprovoked, charged a native workman en route to his job. The attack was witnessed by his companions and the white boss. The attacked man threw himself to the ground and only the gang's immediate yelling rush at the still jabbing buck saved its victim from injury or death.

These instances of attack are certainly rare and should be attributed to special circumstances. As a small boy armed only with an air rifle, when I collected birds in and around the bush and in the edging grasslands occupied by these bucks, I was never attacked. Even more recently although armed only with a camera or tape recorder I felt no concern in similar areas.

Whatever the characteristics of the ram, the bushbuck *doe* is certainly timid, nonaggressive, and almost always the first to leave a patch of cover which is being hunted. She is conspicuously red with numerous white stripes and spots; the ram is often dark, almost black when seen from a distance, with only a few obscure reddish or white obliterative markings. Bushbucks are among the most difficult antelopes to see as long as they remain in their natural environment, the bush. In the open the brightly colored doe is more easily noticed than the ram. Frequently, when a doe is seen grazing at the edge of the bush, the dark-bodied male escapes attention although he may be only a few yards away. Often he is detected only after a shot has been fired and he dashes off to shelter.

Generally, it is the female of animal species rather than the male that is dull-colored. There are, of course, exceptions to this rule, especially in those species in which the male assumes the responsibility for family care. It is difficult to explain why the procryptic coloration of the bushbuck appears to be reversed. In an attempt to find an explanation for this seemingly anomalous coloration, I have spent many hours along the delightfully shaded banks of the Hluhluwe River watching these antelopes, without being able to find an answer.

On one uncomfortably hot day in the 'twenties, I observed a clear-cut case of protective coloration in bushbuck on the flats back from the river edge. The air was vibrating so that the trees and bushes looked distorted, and most animals seemed to have moved into cooler spots near the water. Continued hiking had lost its attraction for me and absence of animal life elsewhere made the tree-shaded river bank seem the logical place for rest or quiet observation.

After looking over the ground carefully, aided by a Zulu, and having seen nothing, I rested my head against a tree trunk and dreamily gazed at the opposite bank. The heat made me inclined to doze, but after a few minutes the flick of something white, fifty yards away, exactly where I had been steadily gazing, re-vealed a bushbuck doe in the open, almost in plain sight. I did not know how long she had been there, but since there was no place to hide on either side, she could not have walked to the spot unnoticed. Probably she had been standing quietly, and only the flicker of her white tail had brought her to view—an illustration of the well-known fact that any movement defeats even excellent concealing coloration.

I was anxious to obtain a good specimen of a male bushbuck and carefully scrutinized the ground in her immediate vicinity in the hope of seeing a ram accompanying her. I pointed out the doe to the boy and told him to look for the male, when suddenly two or three paces to her right there was a slight movement, an ear flick this time, and standing in plain sight was a full-grown ram with a nice pair of horns. He had unquestionably been standing there all the time, dozing in the scanty shade, inadequately concealed, about a third of his body behind a scraggly branch of thorn tree.

A single shot killed the ram. The noise startled the doe, but she failed to locate immediately the source of the sound and stood, looking about, trying to decide which way to run until, when we moved, she saw where the danger lay and dashed toward the bush. Beside her and almost in plain sight a second and still larger ram appeared as though by magic, and both made off into shelter.

On many other occasions in slightly different settings, I observed this reversal of protective coloration—evidence that the male bushbuck is better concealed, at least from human vision.

Until his horns are several inches long, the young male has almost the same markings as the female. These markings are retained sometimes, at least in the East African variety, until the horns are eight or nine inches long; yet individuals in the red phase, though obviously older and presumably more experienced than the smaller-horned black ones, are more frequently shot. This, together with other evidence, seems to indicate that the dark animals are the more effectively camouflaged. These conclusions are, of course, based on man's observation and eyesight but the leopard rather than man probably always has been the most definitive factor in the life of the bushbucks; the question therefore arises as to what colors a leopard sees. The bushbuck offers a profitable and enjoyable opportunity for ecological studies because of its beauty and grace, the charm of its surroundings, and its persistence in an area long after most other game has been killed out.

Zebra

One large African mammal especially has been the subject for discussion by students of concealing coloration—the zebra. Many capable naturalists, including Theodore Roosevelt, most of whom observed them on the nearly treeless plains of East Africa and other open districts, claim that the markings, rather than serving as protective coloration, are distinctly disadvantageous.

In the light of articles by opponents and proponents of the theory of protective coloration, it was interesting to make observations at first hand in the Zululand bush country and see that, as is often true, under different conditions there may be different answers. One conclusion might result from observations on the open plains of East Africa; another from observations in the scattered bush of South Africa. To evaluate both it would be necessary to know the climatic and especially the vegetative history that has been associated with the evolution of the zebra. It may be that zebra striping represents a relic pattern surviving from times when Africa was primarily covered by thin forests.

In the open areas of the bush and on the plains where zebras are seen from a distance, the dark-and-light bandings merge together to form a light gray or white color that can be seen very readily. Under these conditions the only animals that are more conspicuous than zebras are those that are dark, for instance, wildebeest and topi antelopes. At close range, however, zebras are quite different—the bandings of color become distinct and sharply contrasted, so that zebras would seem to be even more conspicuous than the other animals.

Zebras are probably the first game animals a visitor will see in the Zululand reserves. In the open on distant hills they will appear as light-colored dots shining as conspicuously as on the East African plains. This was always true when I found them grazing in the open. In such a situation their color in the daytime is certainly not protective, especially as they go in herds. They are usually aware of an intruder's presence when he is still half a mile away, and they keep close watch on him. In open country this is easy, for their head-high stance when alert is an advantage over most enemies including the lion. Until "recently," that is, until the bow and arrow, man has not been a

sufficiently effective enemy to influence the evolution of concealing coloration. Allowance for this factor must be made when
any theory concerning the relative values of color, form, and behavior in the evolution of African animals is formulated.

Zebras may allow a lion or man to approach to within two
hundred yards. In order to understand the reason for this seeming fearlessness one should remember that in grass two or three
feet high these gregarious animals, forming in a sense a multi-
eyed organism, have every visual advantage over the much lower
lion or virtually weaponless man also hunting by sight. The many
eyes of the zebra herd look down into the grass from a high
vantage point, and, once the danger is seen, the contest immediately becomes one of keen judgment of safe distance and
speed. On the other hand, under these conditions there is practically no protection by this coloration, and the battle resolves
itself into one of alertness and eyesight.

In the Hluhluwe reserve it is possible from some high points
to look down into the valley and with field glasses watch the
zebras among the scattered thorn trees. The preconceived notion
of discovering them by their light stripes is wholly disproved
here. Only now and then in the more open areas will eyes so
preconditioned pick out individuals, but in the bush it is almost
impossible to see them unless they are moving. Time and again,
seemingly every inch of the plain dotted with thorn trees can
be examined without result, because during the day the animals
often rest under tree cover, and although the trees may not be
sufficiently thick to hide them by sheer opacity, they do cast a
dappled or spotted shade. As a test for the presence of zebras,
a rifle shot will almost instantly bring several scattered herds
to view, their motion and emergence to the open revealing them
immediately.

On one occasion from the summit of Mahlungulu I saw a
few zebras—I counted six of them over a period of time—drinking at the edge of the Hluhluwe River a few hundred feet below and a quarter of a mile away. One would drink his fill and
slowly move into the sparse shade of a large, flat-topped thorn

tree, where it would almost immediately vanish. By directing the glasses steadily at one individual, I could follow it into the shade, but if I glanced away, even for a moment, I lost it.

For half an hour this continued, until finally two rhinoceros wandered down to the river to drink. The zebras at the drinking place moved out of the way, evidently quite aware of the brutes' tendency to inexplicable bursts of truculence, but they moved only six or seven yards from the rhinos' path before turning idly to watch the intruders. After a few minutes of drinking, the rhinos turned lumberingly, apparently ignoring the zebras so circumspectly making way for them, and headed for the shade of the big thorn tree. To my surprise, from the place that I had so carefully examined, came a herd of *thirteen* zebras that had been concealed in that bit of shade. Such experiences were not unusual, and they left me with the impression that the bizarre markings of the zebras are distinctly protective under certain circumstances.

The African's Skin

Many times in the 'twenties and later, when I observed large animals and thought about the adaptive and evolutionary significance of concealing coloration I also reflected on the fact that the black man of Africa in the bush indubitably displays concealing coloration of remarkable effectiveness. Every one who has lived and hunted in the bush with the Africans has realized how difficult it is to see them in dim light. Although it is not generally admitted by anthropologists, from the naturalists' viewpoint the black African fits the well-known biological picture for concealing coloration. Other explanations for dark pigmentation, such as protection from heat and ultraviolet light, have been advanced, but comparatively recent information indicates that absorption of long-wave (heat) radiation is much greater in dark than light skins, and that therefore black pigment itself adds materially to the heat load. As to damage by ultraviolet light, protection against these rays comes chiefly from the thickness of a transparent surface layer (the *stratum corneum*), rather than from the underlying color. For these reasons some other explanation is needed for the very dark color that characterizes Africans and so many other tropical people.

Because we cannot experiment with man, the problem of the meaning and origin of skin color in man may be served best by comparison with homologous situations in other organisms. From the natural-history standpoint, tropical peoples' pigmentation, which varies from almost black to lighter shades, more logically may result from natural selection operating in favor of the concealing effects of dark-skinnedness, hence against revealing whiteness, than from any other factor that has been proposed up to the present. In general the degree of color saturation in man is in accord with Dollo's law—which states that light-colored animals are found in the driest areas with a gradation to dark animals in humid lands—to about the same extent as seen in other organisms.

From the natural-history standpoint the pigmentation of tropical peoples qualifies them as examples of Dollo's law about as well as many other random examples of this principle.

In order to clear the way for this digression from a description of wild animals to an attempted explanation of human skin color, still another digression is required. In extenuation of these departures from the major theme, it is possible only to suggest that biologists in general consider man as one of the most important elements in a fauna, and following this conviction we need to understand him fully as well as we do the other organisms.

There are ample technical reasons for restating Dollo's principle in terms of the lightness or darkness of an animal's habitat, rather than of humidity per se. Except in semiarctic conditions, increased humidity and rainfall always, or almost always, produce more and closer-set trees, hence more widespread shade, hence darker soil (humus, specially damp humus) as well as fewer bright cloudless days to begin with. Strongly reinforcing the relationship of an animal's color to the light reflected from its habitat (rather than humidity itself) or the incident light, is the fact that in these very areas of high humidity and dark pigmentation, the resident fauna of the foliage canopy is almost without exception suffused with brilliant greens and is very far from being dark colored—in fact it is notoriously brilliant. By contrast, and at the same elevation, the twig and trunk residents and the forest-floor fauna (where man also lives!) are very dark, that is, they are "saturated" with dark pigment.

When we consider the opposite end of the scale (where animals become lighter in areas of decreasing humidity and brightest in most arid regions), we find usually that black does diminish, except where black lava covers the ground, or where the unshaded native earth is strongly colored. On deserts the most sedentary surface dwelling diurnal and even crepuscular animals, the reptiles, will match their background color irrespective of any other factor including high temperatures and extremely low humidities. On the huge expanses of black undecomposed lava in deserts, the reptiles (and sometimes some of the permanently resident mammals and a few birds) are black or nearly so. A few yards away, on patches of colored ground that are sufficiently large and can support a resident population, if the soil is red, or yellow, or white, or magenta, the diurnal, surface-dwelling, indigenous, permament, and nonmobile residents will match the dominant color of their environment. Obviously, the physiological advantages of albedo (the degree of reflection or absorption) are overwhelmed by the necessity for being concealed. It is even unnecessary to visit African deserts to witness this phenomenon— our southwestern deserts present many classical examples of color matching.

So far as coloration is concerned, natural selection apparently operates most harshly on animals that are most conspicuously colored—that is, those that deviate from the type of concealment provided by their environment. However, this does not mean, as we so often assume, that the harsh treatment is meted out by predators only. Most animals discussed earlier are partially or wholly predators themselves, and stalking their prey or successfully lying in wait for animals to get within their range, requires inconspicuousness or concealment. The alternative to matching background may be starvation—or much longer exposures while attempting to get food—and a greater chance that their own predatory nemesis will get them.

So far as animal coloration is determined by elements of pressure from the environment, concealment comes first and is the definitive factor, for animals that are diurnal, that do not stray from their environment, and that have either originated under favoring color conditions or have occupied a color-defining area long enough to have been shaped to its needs—or to have had the mismatches eliminated. There are numerous examples of this matching of backgrounds, in fact there are few exceptions to the rule, and these exceptions are explainable by other phenomena. Apparently a sedentary nature and daytime activities are important in the development of matching color, but the examples range through such diverse creatures as insects, mollusks, reptiles, and—less frequently because of greater mobility—birds and mammals.

If we are to judge man's biological attributes on a biological basis rather than to treat him as an object of special creation and thus beyond the laws or principles of nature, in what respects does the dark-skinned person fit into the pattern of nature as we have seen it in other creatures?

Certainly there was a time when man was weaponless, and had no knowledge of the use of fire. During that time he gathered his food as he could find it, mostly on the ground if anatomical evidence can be trusted, but he could climb trees for fruit (and he still does). When man was intent on food-getting, inconspicuousness was a valuable protection from his enemies, but if he needed animal protein, he also needed concealment from his prey because he possessed no weapons. However, hand-held stones, later sharpened stones, stones supplanting the tip of a rather dull-pointed stick, and still later throwing weapons that could reach farther and necessitate ever-lessening need for getting within leaping or arm's length distance, diminished some of the need for concealment and increasingly emancipated him from control by the nonhuman elements of his environment.

As he achieved each step in weapons, he should have attained greater ecological independence and supposedly would have been free to invade and survive in increasingly different environments.

It would be interesting to discover the degree to which domestication of dogs enhanced man's capacity to survive the numerous dangers attendant on a primitive way of life. Man's sense of smell is very poor in comparison with that of his dogs, his night-time vision is inferior, and even his auditory sense seems to suffer by comparison. The human being, of today at least, is overwhelmingly sight-dependent when compared with his canines. When primitively armed man domesticated a small carnivorous animal that was so notably endowed in a sensory area where he himself was so poor, he gained an immensely important supplement to his total effectiveness. Simultaneously he reduced the value of concealing coloration for himself. From that day to this, dogs and man have maintained an extraordinarily effective partnership in procuring food and avoiding surprise attacks.

With distance-reaching weapons and fire for protection at night and fire for aggression by day, man should have become relatively free from all predator danger while at the same time he was becoming a far more effective predator himself. There would seemingly have been far less need for concealment, and although it does not necessarily mean an automatic divergence in color for each new habitat, such mutations as might be unsuitable would not be speedily eliminated. However, the scarcity of favorable mutations, especially of the few skin-color genes, makes their occurrence almost impossible in terms of statistical probability, and more improbable that changes could be meaningful in an animal that had gained ascendency over its environment. With complete mastery of the environment any later mutant genetic color change might go in any direction that accident directed.

However, intraspecific conflict, in contrast with inter- or extra-specific relationships, has probably characterized man since his preanthropoid existence. Interfamily, interclan, or intertribal battles for resources (territoriality conflicts) or for mates can be a logical assumption (we seem not to have lost these proclivities), especially in hot, humid climates where ornamentation has a higher priority than body covering even now; there victory would go to the best-concealed warriors who could approach or lie in wait for their enemies and strike with surprise. If, armed with primitive weapons, a force of naked whites met an equal force of naked blacks in the forest or bush, in a battle to the death, the blacks would win with at least a 50 per cent larger number of survivors. Only the Romans could stage such a battle and this experiment would be the only method by which we could obtain the experimental data we need.

Despite the absence of experimental evidence, I am certain that no naturalist trained in observing the phenomenon of concealing coloration can escape noting the simple fact of the remarkable effectiveness of dark skins in achieving near-invisibility in the bush or forest; in fact, during the Second World War, our troops in the jungles of the Pacific found it more difficult to detect the Japanese (whose skins are relatively dark) than to see the faces of white troops; this object lesson was not lost on either the Japanese or our men, both of whom soon began darkening their skins with charcoal or mud. Certainly these men quickly recognized the effectiveness of black skins as a practical form of "natural selection," even though they may never have heard of procrypsis and selection in the evolutionary sense.

This long digression from Hluhluwe game animals cannot be brought to a conclusion until a few other comments on alternative theories are touched upon.

The alleged capacity of the black man better to survive and to do physical work in hot climates might be more correctly explained as rationalization in support of slavery, while the white supervisor stands in whatever shade is available. Because he is largely limited to employment in physical work, the average Negro is often in better physical condition than the white man to undertake strenuous exertion, and this superior conditioning helps support the myth that his coloration is a physiological rather than a concealing adaptation to tropical conditions.

When the African Negro's work habits are governed by his own choice or conditions, he elects to avoid the heat of the day and works from dawn to midmorning and again in the late afternoon. During the hot mid-day he is just as prone to rest in the shade as any other human being, and with his philosophy and ancient custom he does this with little fretting or worry.

It has often been argued that the other hairless mammals of

Africa are dark-colored, and that this may be indicative of the need for pigmentation as a necessary protection against actinic light. The rhinoceros, elephant, hippopotamus, and the almost hairless adult African or Cape buffalo are the examples cited, but all these animals are special cases, exceptional for the fact that they are virtually immune to effectual attacks by predators. They all repair to the shade when the weather becomes too hot, or they take mud baths, or seek shelter in water. Their enormously thick skins, their dominating size, and their evolutionary history render them poor examples to cite in support of any theory concerning the reason for their present coloration. Nonetheless it is remarkable how effectively they do achieve concealment, from man at least, through their adventitious coloration. Because of their neutral color and their habit of wallowing in local mud holes, the rhinos and elephants often assume the natural color of the soil around them; and because of their virtual immobility when in danger, they become near-mimics of boulders or of the innumerable large termite hills so characteristic of Africa.

Because of their relative immunity to attack by predators, this concealment was probably immaterial to their welfare until the comparatively recent advent of man with firearms, and their color is therefore not especially significant in connection with man although it is an interesting fact in itself. Their mud packs and love of shade and water may be more logically associated with such factors as size, protection from biting insects, heat regulation, and concealment through immobility, than with any single environmental factor.

When confronted with the overwhelming evidence for concealing coloration as it has evolved in so many other lines of animal life—insects, birds, reptiles, and mammals—a naturalist, working with Africans in the bush, finds it easy to assume that a white skin would be a disadvantage and a black skin would be selected for its fitness for survival as a concealing device, even far back in time and the evolution of man.

In view of the concealing value of black skin in shadowed country and black or brown naked facial skin of the primates, including the anthropoids, it is difficult not to believe that man originally may have been "black"; if so, white skins, rather than black, require explanation. Certainly, white would seem to be a freak of mutation of little survival value per se even in far northern areas. On the basis of color I would hazard a guess that man originated in well-wooded or densely-forested country.

Since early childhood I have repeatedly observed the effectiveness of concealing skin color. But scientific interpretation requires more knowledge than a vivid recollection of the near-invisibility of an African in the bush or after dusk at the edge of a clearing.

There is another, an African, explanation of black skin color which also takes note of the contrasting whiteness of the palms and soles. Whether or not this is characteristically Zulu I cannot be sure, but all Zulu small boys, and some that are not so small, delight in making clay models of animals and people and setting them in the sun to dry. The black man, according to legend, has white palms and soles because when God had made original man he set the models out to dry in the sun in a quadrupedal position, hence the undarkened hidden surfaces of the palms of the hands and soles of the feet. Among other things this legend suggests the recognition of the tanning action of the sun even on a black skin.

In Africa, even today as it was in my youth, man and his fellow animals intermingle in a continual kaleidoscopic mélange of activities and relationships, and by all odds the happiest contacts with either are at the level of the simple aesthetic, nonanalytical day-by-day contacts.

In the 'twenties, much as during my second return trip in the 'fifties, day after day in the shooting areas and in the Hluhluwe reserve itself, I jotted down observations on animals, but also on the Zulus and their way of life—observations so biological in nature as to justify them as legitimate natural history even though anthropological in a sense.

After each day of arduous physical exertion I would start the long trip back to camp. But despite the fatigue, the weary tramp homeward under the setting sun was always joyous. In the winter months the brown hills turned purplish red in the slanting rays of the low afternoon sun. All day the distant hills may have been hot and hazy blue, but in the last glory of sunset, the blues became purple, violet, and rose.

The colors would hold for only a few minutes, touching the hills and the valley until all the world seemed to glow with warm color; then as the sun dropped lower still, the valley filled with smoky shadows, pools of twilight in the fading day. Last of all, the hilltops glowed like giant opals in a smoke-gray setting, then the color faded, and the sky, for a few brief moments, reflected the colors that were on the hills.

The chill of evening would fall, and in the twilight while the dew-wet grass was fragrant, I would walk along the hill until down below the faint glow of the campfire would show the homeward way, making one spot of welcome light in the blackness of gathering night.

3

RHINO CAMERA HUNTS IN THE 'TWENTIES

In the 'twenties, photography in the Hluhluwe reserve was always a delight and often exciting. I liked to visit there in the winter, in many respects the best season for photographing the wild animals. Usually the air was cold and invigorating, just right for enjoying to the utmost the hours of walking and, when tired, the pleasure of stretching out in the tall grass, protected from the breeze, but exposed to the warm sunshine.

In the morning the air would be fragrant with the scents of a new fresh day—the simple primordial perfume of growing things and the open country. Under the glowing early light, the wide fields of grass would give off faint aromas, such as the delicately acrid but spicy scent of bruised thatch grass, *isicunga,* or the faintly smoky, aromatic odor of *umbutambutane* grass.

Natal's winter coolness is probably surprising to Americans; it should attract tourists during our summer vacation—South African winter. And the winter colors, although generally pastel, should lure color photographers. Most of the grass in the open areas, the parklike savannah country, is reddish brown with fine purplish streaks on the blades and a reddish color on the stem. At a distance, in the flat light of noon, it may look brown or golden depending on subtle differences in the atmosphere, but with the changing angles of light and greater atmospheric filtration of sunrise and sunset, the colors are exceedingly variable.

51

It was under these colorful conditions that I enjoyed big-game photography some thirty years ago.

One of my longest and most interesting rhino camera hunts started on a morning when I studied the Hluhluwe valley on both sides from the crest of a ridge near one of the camp sites. I looked through my field glasses at some suspicious appearing gray termite hillocks which so often suggest rhinos. As usual I was accompanied by one boy carrying my gun, one boy carrying the camera, and the native game guard. Suddenly one of the boys loudly whispered, *"Nanse u'bejane n'kos"* (There is a black rhino, chief). I replied, *"Ehe,"* implying casually that I had already seen it. The boys talked softly, but agitatedly, and I wondered at these men becoming so unusually excited over the presence of a rhino half a mile away. I placed the glasses back into their case and we started off, trying to get close to the animal.

After searching awhile in the dense bush I suddenly saw it only twenty-five yards away, half hidden in the tall grass and short thorn bush—squat and brownish, on its nearer side a long, curved horn and a black, piggish eye that stared unblinkingly.

It was impossible to get into position for a better picture because the grass was long and obscured the view, but, though partly hidden, the rhino presented a fairly good picture, and I made an exposure. At the click of the camera, her ears pricked up more alertly, and the big form seemed to surge forward a few feet. The motion of the legs was hidden, and the heavy porcine body showed only this forward progress. It gave the impression of an enormous floating weight, with an irresistible strength and force inside it.

I took a second picture. Now the huge animal turned once or twice as though undecided, the head swung threateningly from side to side. Then, while turning, she revealed a calf, half out of sight behind her huge body, and they both trotted off into the brush. Although we had not been too close, we felt relieved when she had gone. Our morning encounter was informative, but unadventurous even though she was with her calf.

Meeting these unpredictable animals, which can be decidedly

dangerous at times, always evoked in me the psychological elements present when one unexpectedly runs into a strange dog in
the path. If the dog stands his ground and raises his hackles,
safety may be gained by either of two dignified procedures: by
attempting a nonchalant and inconspicuous retreat that will not
seem like flight and so not encourage attack, or by facing the dog
and trusting that his morale will fade. To imagine my feeling of
tension in meeting a rhino one merely needs to visualize the
situation when the dog weighs several thousand pounds and
when the surrounding grass and brush are tall and entangling
enough to hold up a man even if he runs for his life.

These often experienced moments of tension made it evident
to me that the evolution of man must have been achieved by
many compromises, and that man's erect posture, no matter how
valuable in some circumstances, was by no means the product of
fast escape in tall vegetation. Even the quadrupedal rats and
small antelopes could speed through these tangles with far greater
celerity than longer-legged bipedal man. In fact, whether motivated by intellect or instinct, I found myself appraising any
nearby tree for climbability as well as stoutness under the possible impact of a rhino. I wondered sometimes if man's kinship
with the arboreal primates was as remote as he had thought it to
be. Certainly many men have saved their lives by emulating their
ancestors. The compromises retained by evolutionary processes
do have their value, and primitive man, lacking firearms, still employs simian tactics with profit. As judged by man's aptitude for
employing this escape device, it seems as though it must have
been of selective advantage up to an evolutionary yesterday.

With increased experience one learns that, formidable as it is,
the rhinoceros can be photographed with comparative safety, but
at first a man strolling through big-game country with only a
camera and no rifle at hand, feels extraordinarily naked and helpless. It is not surprising that one's resolve not to carry a gun will
suffer a fate like that of the monthly resolution to quit smoking.

After the female rhino and her calf had removed themselves to
a safe distance, we continued our walk down the ridge. Soon a
breathless messenger announced that he had seen another rhinoc-

eros and calf. As fast as tall grass and thorny vines permitted, we dashed off the few hundred paces to a spot where a waiting Zulu pointed out the animal as it stood motionless in the tall grass across the valley about one hundred and fifty yards away. She was a cow with unusually long horns, and although in the 'twenties no white rhinos were supposed to occur in the Hluhluwe reserve, one boy insisted that she was an *'umkombe,* the extremely rare "white" species, while another was equally emphatic that it was the *'ubejane,* the "black" rhinoceros. The argument was never settled, for in the concealing brush and tall grass only the back was at this time clearly in view, and even later we never came close enough to tell for certain. With experience it is easy to distinguish the so-called white rhinoceros with its distinctive back outline and differently shaped head, but as none of the boys seemed to be able to tell the difference, and the animal was so wary as to prevent our getting a good look, we accepted the odds against her being the docile and safe-to-approach white species and treated her with circumspection.

She was in country difficult to photograph, but it was worth the attempt because of the calf which could be glimpsed from time to time.

Between us lay the narrow valley, which was filled with tall brush, five-feet-high grass, thorn scrub, and a few palms. It was bad country in which to avoid a charge, and under those conditions even a good marksman would have found it almost impossible to shoot and deflect the animal until too late. Also, if it turned out that we were dealing with the precious white rhino, no set of photographs could have justified endangering the animal by shooting at her in self-defense, should she have charged when frightened by the click of the camera.

There was practically no breeze. In the tall grass and brush I would be able to conceal myself perfectly because of a rhino's eyesight, which is poor in either species, and I therefore decided to approach as closely as possible and then climb a tree to get an unobstructed view for the camera. This seemed to be the only possible method by which a good picture could be obtained, and at the same time, if I could reach the tree before a charge, there would be safety for both man and beast. From the ground it would have been necessary to approach within five yards of the animal and her calf, and I could not persuade any of the boys to do that.

I ordered the boy carrying the bulky camera case and extra gear to wait in a safe place. Then I began a careful creeping stalk down the hill, into the tangle of bush, and the walk across the small valley. As soon as I had passed the worst of the thickets screening the cow I focused all attention on her movements.

The tick birds, the worst avian pests of the rhinoceros photographer, were alarmed and "chirred" almost continuously. The cow was alert, but not knowing exactly where the danger lay, surged first in one direction and then another. She acted as though she had received some whiff of tainted air, but not enough to provide a decisive stimulus. The varying breeze made continued stalking impossible. At times it looked as though she would break away and move over the ridge behind her, then again it seemed as though she might rush down the hill to investigate.

It was while she was thus occupying the center of attention, when my every nerve was tense with expectation, that there was a sudden explosive snort and crash in the bush a few yards away and a black rhinoceros bull emerged, standing seemingly at a distance of only a few arm lengths and truculently staring back toward the spot from which had come some sound or smell.

He had evidently been lying down, asleep in the tall brush and grass, and I had walked almost on top of him. He was now standing only thirteen yards away, as we measured later. There was a deep, breathless silence, and we cautiously looked about, searching for the nearest tree. I hardly dared to move even my head for fear that any motion, no matter how slight, might attract the attention of the mammalian tank.

There was nothing to do but wait for some overt act, and the rhino evidently considered himself in the same predicament. He could not make out the source of danger, and fortunately no turn of the breeze relieved him of his indecision.

For many long minutes there was absolute silence, and I was concerned lest at any second the fickle breeze might change.

There was no telling what might happen, for in bush so thick the air does not move in straight lines but eddies and seeps here and there in vague and unpredictable vacillations. At such a short distance the scent might have carried to him at any moment. If he knew that the danger was almost beside him, it seemed probable that he might make a rush up that trail of scent.

After what seemed an eternity his head began to turn, first to one side, then to the other, his eyes short-sightedly staring with piggish intensity and examining every suspicious object within range. With ears pricked and all alert, he slowly turned, moved a few hesitating steps, stopped to look back, moved again and, suddenly, as though puzzled, trotted off, looking back first to one side and then to the other as he sidled along. Each time he turned, he looked ridiculously like a dog that had been bluffed into retreat and expected to be pursued. He appeared uncertain whether to reverse and speed up in ignominious retreat or to stand and face the danger. Finally he panicked and plowed away through the tangle of brush at full speed ahead, knocking down a small, thin tree that stood in the way. When he disappeared over the ridge with a crackling of brush, we breathed deeply.

Now we remembered the cow on the hillside fifty yards away. She was still standing motionless, staring toward the vanishing sound of the retreating bull as though in a daydream. That was her only reaction, and doubtless if we had not been present, she would have continued to stare until she would have passed into a doze, and then a sound sleep. As soon as we started moving, however, the tick birds commenced their chatter and she turned and faced directly toward us. There was no doubt that we were well beyond her range of vision, especially since we were up to our necks in tall grass and brush. There was now practically no breeze, but what little there was would probably move up or down the valley according to thermal differences rather than across it, and it was unlikely that our scent carried to her. Most animals when alarmed turn and look in all directions, and do not seem to have any "uncanny sense" of where danger lies. But the rhinoceros cannot bend its neck gracefully and look back over its shoulder—it is constructed too solidly for such an acrobatic feat.

When danger threatens, the animal turns stolidly and faces toward it—a most uncompromising attitude that, justifiably or not, enhances the rhino's reputation for truculence—and automatically aims its head in such fashion that when panicked the rhino gives the illusion of a charge.

About twenty-five yards from the cow was a slender mimosa tree that looked as though it would afford a sufficiently secure perch. I carried the camera six or eight feet from the ground, where a forked limb and some small branches gave just sufficient support to allow picture taking if the rhino moved into the open. With each snapped twig, the cow moved forward a few steps, then, hearing nothing more she moved back again. The tick birds continued their alarms, she became more and more nervous, finally moved off to the right, and disappeared into thick brush frustrating our picture-taking efforts. During all this time the calf had remained hidden behind her.

Before we continue our photographic hunt on rhinos, a few comments on the tick birds might be inserted. Although these starlings, known to the Zulus as *umhlalanyati* ("sit on the buffalo"), help their hosts by passing the alarm to them, their aid is supplemented by an apparently keen sense of smell and hearing in the rhino. On the other hand, it soon becomes obvious that a rhino is almost blind, at least for the interpretation of motionless objects at even less than fifty yards, and for this reason one is amazed to see how quickly it sometimes seems to sense the direction of danger even in still air.

Sometimes when a man has only a part of his face exposed from behind a tree, one rhino may turn and stare in his direction; another, however, may fail to see a motionless person only half the distance away and in plain sight. Coincidence might be the explanation, but the response seems to be too frequent to be accidental. The real facts may be simpler despite those adventure writers who endow wild animals with mystical senses.

Tick birds feed on all sides of an animal, and a flock of half a dozen or more settle on their host, scatter over its body, and begin a search for ticks—not a difficult undertaking in Africa. As soon as they sense a threat, they immediately call their alarm notes and retreat to the side farthest from danger, where they cling almost out of sight. When alerted they usually manage to keep their host's body between themselves and the peril. The few lookouts that keep the enemy in sight show only a small part of their body, and the closer the danger, the less of their bodies shows. This habit of retreat to the side farthest from the threat could quite obviously provide their host with cutaneous information as to the direction of danger, and this might be reinforced audibly by the bird's habit of uttering almost continuous warning sounds. The possible effectiveness of this type of mutualism is enhanced in animals that are accustomed to face toward the source of danger, the rhinoceros and buffalo particularly, for under these conditions the host animal swings about in varying directions until it locates the source of a disturbance, and while doing so the birds usually scramble hurriedly from side to side, clinging with their powerful feet and needle-sharp claws. When they have finally retreated to the rump area and cluster there, calling harsh warnings, it would be a dull animal indeed that could not sense at least the general direction of the unseen menace.

These beautiful little starlings, somewhat reminiscent of the American waxwings in their neatness and sober hues, feed on the blood-engorged ticks that so beset almost all game and domesticated cattle. In the farming country the birds are becoming rare, presumably because of the arsenical dips in which all cattle must be submerged at regular intervals, and it is therefore chiefly in the game refuges that one will have opportunity to study their habits. If their services can be classed as mutualism of a sort, it must be doubly effective, for the birds provide dual benefits to their host—the removal of damaging and irritating ticks and an effective warning against danger.

As the day of our picture hunt was getting warmer, it seemed probable that by going down the bed of a small, almost dry stream, and keeping in the shade of the scrubby trees, chances for finding another rhino would be good. I was not too enthusiastic about working in the dense scrub cover because of the possibility of an inadvertent approach to danger and because of the obstacles to taking good pictures caused by the screen of brush. But we would frequently find open spots in the brush, usually the dust or mud wallows of wildebeest, where "candid camera" tactics might produce results, and we hoped that with rhino so abundant, we might find a sleeping animal more or less in the open at the edge or center of one of these openings.

The valley through which we moved was a peaceful looking spot, beautiful with clumps and groves of African sandalwood and wild date palm.

The palms were *isundu* (*Phoenix reclinata*) from which the local natives make palm wine. They cut off the leafy tops, trim the tip of the trunk in such a way as to drain the sap into receptacles, and after two days the sap is fermented sufficiently to be called wine.

Signs of rhinoceros were abundant in all directions—tracks as big as dinner plates and chewed thorn-tree twigs, their favorite food.

After some time we reached the banks of the Hluhluwe River, and after the hot and exciting work of the morning, it was a relief to wash and drink. The Hluhluwe of the 'twenties was beautiful and clear, running over stony riffles, through long, reed-margined pools, over sandy stretches where the water sparkled in the sun, then through thick, black bush with hardly any perceptible flow and the pools looking deep and inky black. Although the river appeared so inviting, and one longed to plunge in for a pleasant swim, it would have been almost suicidal to do so. It would have been dangerous even to wash at the margin of the deeper pools. The Hluhluwe River was, and still is, a favorite haunt of crocodiles, many of them large, and all hungry. Each year the natives suffer accidents from crocodiles. However, with the passing years, erosion, rapid run-off and resultant reduction of underground water are gradually altering this river, and in time the crocodile pools and their occupants will become rarer and eventually may disappear.

After a rest and a light snack, we returned by a different route, through more open country than that traversed in the morning. We walked up-hill for a quarter of a mile and soon found ourselves on the top of a ridge from which we could see for miles. Below us ran the silvery Hluhluwe, and from it in all directions wound the sinuous game trails. Most of them seemed to converge at definite points, showing that many animals preferred drinking at shallow and well-trampled fords, rather than at the deeper pools in more isolated spots.

And now our search was rewarded. From where we stood we saw a rhinoceros lying asleep. It was on the far side of the valley, and with the exception of a small open space immediately around it, the ground was thickly grown with thorn trees six or seven feet high, so that it was necessary to detour to one side, walk blindly through the high brush, attempt to recognize a larger tree used as a landmark, and then to reach it without disturbing the sleeping animal. After half an hour of anxious stalking, we finally reached the tree, only to learn that the rhinoceros had been awakened, the garrulous tick birds undoubtedly being responsible. But he had not left the spot and seemed drowsy.

There was no chance of getting pictures from the ground, and so I climbed an advantageously situated thorn tree. Even from the main fork, eight feet from the ground, the view was so poor that it was necessary to force a way through the interwoven thorny crown of the tree. Fortunately, my heavy pith sun helmet made a good, though noisy, battering ram, and by pushing through the branches and thorns, and carefully squirming through the small opening thus made, it was possible, without too much discomfort, to get on top of the tree crown.

Even the final position was far from comfortable, however, for it was necessary to stand half doubled over, and any attempt to stretch upright was punished from behind by a tough and awkwardly placed branch bristling with sharp thorns. My efforts to brace against the tree trunk resulted in sharp jabs from the inch-long thorns growing there. After making several exposures from this position, I sank back against the main branches trying to ease my aching muscles and then climbed down the tree. The still somnolent rhino seemed to be listening—then eventually turned his ridiculous little tail over to one side, and hastily barged away through the crackling thorn trees and out of ear-shot. At the moment it was impossible not to envy him his impervious hide.

This was the last rhino for the day, and we returned as the sun was sinking.

I recall another picture-hunting day when I started out with boys and a game guard along the same route, but kept to the ridges along the edge of the reserve in the hope that we would see a rhinoceros where the brush was thin, and so give us better chances for good photographs. After a mile we saw two rhinos standing near each other on a ridge across the valley. One was browsing in the open, the other was almost concealed in a thicket, perhaps lying down. Between us the valley was approximately a quarter of a mile wide, and in the center of it lay a narrow strip of bush, called by the local natives *ihlati kwa gube* or *kwa gube* (bush of danger, or evil bush).

The name was unquestionably appropriate. It was a very dense bush and the wild tangles of vines and thorns allowed travel only along the game trails that formed a network of paths, all showing abundant rhino signs. The natives claimed that the bush was inhabited by all kinds of animals, both natural and supernatural. Most men would refuse to enter it, and I was not surprised when the camera boy said that if I insisted on going closer to the rhinos, we should detour this bewitched bush. This would have necessitated going to the head waters at one end of the valley a mile higher up, then crossing to the other side—a two-mile walk to save going through that narrow strip of *kwa gube* bush land. I expected that my twelve-year-old native gun boy would side with the superstitious camera boy, but when I asked him he merely smiled and said, *"Ku lungile"* (It's all right). The native game guard looked scornfully at the big camera carrier and made some caustic comment about a "missing liver," the Zulu expression for faintheartedness.

We started toward the bush, and all went well until we reached the edge. Then the camera boy insisted that he was not going through that bush, and that we could wait for him on the other side. The game guard made some belittling suggestions that eventually convinced the camera carrier that it was better to fall in with the majority, but he was obviously frightened. Perhaps it would be more true to say that he was either exceptionally imaginative or realistic.

While we were going through this strip of bush, the rifle was a consoling thing to have in hand. Game was so abundant that the bush even smelled strongly of animals, and the paths were as well trodden as a shady lane in a crowded pasture. Judging by the tracks, wildebeest were the most abundant animals, but there were many others also, all keeping company in *kwa gube* bush.

When we finally came to the edge of the bush and moved into the open, there were sighs of relief, and we wasted no time in putting additional yardage between us and the edge of that fascinating but dangerously thick patch of bush.

On reaching the top of the hill, we could see nothing at first. The quarry seemed to have disappeared, but after we had walked down the slope of the ridge for a hundred yards, a great, lumpy-looking "ant hill" showed through the bush, and with a little careful maneuvering, the form of a sleeping rhino could be made out. He had lain down in a small, open space. The nearby brush was not more than waist high, which made an ideal place for photography. The other rhino had apparently wandered out of sight.

When we saw the sleeping animal first, he was about a hundred yards distant, and for once the ever-wary tick birds were either asleep or off duty. We cautiously crept toward him, and at thirty yards I motioned the camera boy with the extra lenses and other material into a tree. I wanted him in a safe place, for, if the rhino charged, the boy might leave everything to be demolished. The gun boy took my rifle, and I ordered him to keep about four feet behind us and stay at that distance no matter what happened. Prepared for work, we advanced, trying to keep a small patch of brush between us and the rhino. This brush was twenty-five yards or less from him and made a convenient blind from which to take pictures.

We approached this patch very cautiously, walking directly in line with another and larger patch. When this nearer bush was reached, we slowly raised our heads to see what the rhinoceros was doing. We had been so careful, and had moved so quietly, that he was still unaware of us. We raised our heads inch by inch, almost holding our breath, afraid that at any instant we might hear the tick birds' warning calls or stumble on the second rhino, whose whereabouts we still did not know.

As our heads reached the top of the brush and we looked over it, there was an explosive snort and rush, seemingly from under our feet. For an instant we were paralyzed, then realized that we had almost stepped on four wart hogs, not a rhino as we had thought. During that first split second of the rush, we had expected to see the second rhino confront us at arm's length.

As soon as we recognized the cause of the commotion, our heart beats dropped to normal, and we laughed with relief and amusement at the wart hogs' panic-stricken rush. They made a headlong dash straight toward the sleeping rhino. As they reached him, startled tick birds flew screeching into the air, the rhino jumped to his feet with a grunt, and the hogs swerved off to one side.

The rhino stood motionless as the hogs faced about waiting, and the birds returned to their post. Again there was the same oppressive silence—no sound except the "chirring" of the tick birds and the faint, crackling song of a grasshopper. For a minute nothing moved. Then the wart hogs raised their tails in the air, the brush parallel to the ground like a flying pennant, and rushed madly away. As they went, the rhino turned and gazed after them, the tick birds scrambling over the ridge of his back to hide. Stooping low, we strode as quickly as possible to our chosen bush. Just as we reached our cover, the rhino turned and faced our way again. He appeared to suspect the general direction of some danger, but could not make us out.

I snapped several pictures, and each time the rhino would wheel, sometimes appearing to be on the brink of an attack, at other times simply curious. He would probably have quieted down, but the tick birds kept up their warning calls and scurried from side to side as he turned. Finally, as they worked their way to concealment on his rump, he started toward us, and that was our signal to leave. He was coming slowly; we stooped and left the field as rapidly as possible. Unfortunately, one of us stepped on a twig which snapped sharply, and there was a thumping rush behind us as the rhino broke from a trot and pounded along in our direction.

The gun boy had undergone more than his young nerves could stand and had fled to a tree at the first move; he was still climbing even though he was already twenty feet from the ground. With the advantage of our early and speedy start, it seemed as though it would be easy to sprint to a substantial tree well ahead of the rhino and dodge behind it. The sounds from back of us came nearer. The game guard, who had speeded past me as we started running, dropped back a few paces behind me, and held his gun ready, without shooting. He remembered his instructions well, for I had told him not to shoot unless he must, and yet, of course, he was anxious to demonstrate his skill and bravery. After we had run fifteen or twenty yards, the sounds indicated that the animal had turned. When we looked back he had reversed direction. A last view of the rhino showed a broad, squat stern and a tiny tail disappearing into the bush below us.

I thought that the six exposures made on this animal would probably be good, but I wanted a few more at different angles and with a different background. If I obtained these, rhino photography could be concluded and work started on animals with more predictable dispositions.

So we continued our search. After a while we found the comparatively fresh tracks of a rhino and her calf and followed them along the ridge. As we neared a patch of dense bush we heard a snort. As before I sent the spare equipment to a place of safety. The camera boy soon stowed the equipment in the crotch of a tree and climbed up to the higher branches where he could look over the nearby trees and see beyond us over the bush. He had hardly settled himself when he started waving signals, and I sent the gun boy back to find out what he had seen. The youngster returned at a run to say that just around the next patch of bush there were three adult rhinos, and that a female and her calf were headed toward them. This was wonderful news and walking quietly we followed on the trail of the female.

Just beyond the bush the tracks turned and skirted along the edge of a little island of high-piled boulders from the center of which grew a small tree. This was an ideal spot so we scrambled up and looked over the edge. About twenty-five yards away were four rhinos, one a little smaller than the others but almost adult. The calf we did not see, either then or later. They were all standing in scattered thorn brush, so it was necessary to climb into the small tree to get a better view. The animals were still partially hidden by the tops of the thorn trees, but any picture showing four rhinos together would be a trophy. Slowly, even majestically, the four huge animals moved toward us, keeping abreast, until at eighteen or twenty yards they emerged from the obstructing brush into the open. This was the great moment. I pushed the shutter lever down. There was only a feeble click of broken gears. I tried a second time, but the camera failed again. The nervous strain was great, and my disappointment even greater as the rhinos turned broadside, stood for a moment as though posing, then stampeded down the hill and out of sight.

A few days later as we left camp for the return to civilization, we looked back toward the bush; it was sunrise and the far hills were tipped with lemon-yellow light; in between, the bush still lay in dusky shadow. Close by, an aloe pushed its dull red spikes of flowers into the clear air. The dew was glistening on the grass and on a drop-beaded spiderweb stretched carelessly from aloe bloom to tufts of dark green leaves. The air was cool and clean, fragrant with the smell of morning in the wilds.

More than thirty years had passed by the time I returned, but never during all that time had I forgotten the hues of those mornings, the freshness of the air, the excitement of our stalking the prey. The episodes with wildebeest, bushbuck, zebra, and rhinoceros formed little links of a chain that, in the 'fifties, pulled me back with elemental force to Africa.

4

THE GAME RESERVES REVISITED

And now, on this wintry July afternoon, I was back again in Hluhluwe, after an absence of almost thirty years. Back to those old ridges where I once photographed rhinoceros, back in the once dangerous *kwa gube* bush. But at this time, a generation and a second world war later, I drove comfortably through the reserve on a modern road that followed closely the old haunts.

This land still was beautiful beyond description and beyond the efforts of color photographers who attempt to capture its subtleties. Its charm was the result of a combination of many things, not just of a single feature, such as its legacy of the primitive. There was a feel to this land; and I breathed it in with the fragrance of the earth, the blossoming *umsinsi, umxaba,* and *inhlizio omkulu* trees, the grass fires, and the sun-baked vegetation in this dry weather. Down on the dun-colored soils near the river bottoms the year-round vivid-green tropical foliage of the native palms and fig trees contrasted sharply with the dry hillsides and gave promise of life to come with the rains of spring.

As I drove along in the direction toward one of the modern huts that have taken the place of the old camps, time, at first, seemed not to have moved on in Hluhluwe. The hills were still turning that same familiar milomaise pink at this hour near sunset, as I crossed the boundary line, in the company of a young native boy and the native game guard Panza who might have been the very same Zulus with whom I went hunting thirty years earlier. Yet, things had changed. At one time I had associated this characteristic pink color with a special grass, the fine grazing material known to the Zulus as *insinde,* but now I noticed that this best of all native grasses was practically gone in the reserve. In spite of its disappearance the color remained—the result of a deficiency in soil phosphate as I was to learn.

I saw many animals during that long slow drive to the rest huts, and I noted with a deep feeling of frustration their poor condition. Except for the zebras, wildebeest, buffaloes, and rhinos, most animals were emaciated, some to the point of weakness, especially the impalas; that the zebras were in good shape was not surprising, for, so far as I am aware, almost no one has ever seen a zebra except as a plump, apparently well-fed beauty, with well-rounded hindquarters that seem to denote excellent condition.

As I saw these remnants of long-past days, I was gripped by a sense of loss and loneliness, difficult to describe. It was neither pleasant nor unpleasant, though strongly moving, and it heightened my sense of how greatly privileged I was to have participated in my younger years in a way of life that was so rapidly disappearing. Very soon now the world would no longer provide these experiences that I once accepted as commonplace. The explosive growth of human population would continue its destructive effects, and the next generation would have little or no opportunity to experience an adventure into the age of mammals, except in these mammoth zoos, the refuges.

Things had changed, however, also in a different direction. I immediately noted the abundance of the beautiful inyala antelope, which was almost absent from this reserve in the 'twenties, thirty years ago, although present nearer the coast. Now the inyala was a very common species in Hluhluwe. Impala antelopes also had multiplied prodigiously, but the bushbucks and reed bucks once so common, appeared to have become quite rare.

I wondered whether the multiplication of inyala and impala may have been partly a result of the absence in recent years of an effective predator such as the lion or even primitive man; and whether the enormous increase of monkeys and baboons in the reserves was the result of the near-absence of their once much more frequent predatory enemy, the leopard. But why, then, had the bushbucks and reed bucks diminished in numbers?

The sun had just set when we reached the rest huts, called rondhovels—round mud-walled thatched dwellings resembling somewhat the huts of the aboriginal natives. Here on a ridge, in the heart of the reserve, lives the white game conservator. He assigns native game guards whose duty it is to show tourists the animals they have come to see and protect these white novices from their own foolishness which might cause their own death or that of some valuable wild animal. These government-owned tourist accommodations are available for a nominal fee, and are equipped with community kitchen and staffed with trained servants, who in a separate hut prepare the food the visitors must bring. These conveniences save much time and trouble, if compared with the camps of former days, but the changes are not wholly advantageous. With the new comforts and the shelter of a hut, one becomes insulated from the old feeling of participating in nature. Visual and emotional contacts seemed reduced when I reached the huts. This was especially true at night because the thick roofs and walls muffled the sounds, and a feeling of security lulled the senses. The primitive camps of my earlier trip had provided intimacy and integration with the environment so that for natural-history education they were incomparably superior to these deluxe accommodations.

To get back some of the former feeling of union with the world outside the huts, I went out when evening came and sat under the glow of the sky watching the light fade from its previous incandescence on grass and trees to the usual diaphanous evening haze.

I sat up several hours listening to the many voices of the cool, moonlit night. Among the most delightful sounds were those of the goatsuckers or nightjars, with calls reminiscent of their American relatives, the whippoorwills. From far and near they called, while from the bush at the back of the rest huts now and then came the piteous calls of the galago, the rare lemur or *sinkwe* of the Zulus. From the valley below came the mellowed sound of the hyena, one of the truly beautiful calls of the wild, although it occasionally lapses into hideous, high-pitched squeals and gurglings.

Even with present-day amenities for visitors with their numerous cars, animal life follows its ancient pattern. To see the sanctuary at its best it is still necessary to rise early and to travel while the animals are feeding in the open.

At 5:45 A.M. on the first morning of revisiting Hluhluwe, I was awakened from deep sleep in the comfortable rest-hut beds by my boy bringing early morning tea and a pitcher of hot water for washing. Long before sunrise I was out in the crisp winter air with my Zulu guard Panza.

Our first exploration on foot (five months later tourists were required to stay in their cars except on the rest-house grounds) brought us to a herd of about fifty Cape buffaloes grazing in a woodland park among thorn trees in the half-light before dawn. The animals were so huge and formidable that my old respect for them blossomed suddenly, and, as of old, when unarmed and on foot, I kept furtively glancing about for a good tree to climb. As we walked up to the herd in the predawn morning it became apparent that they were aware of our approach, for they drifted into a crescentic line, bulls, cows, and calves, scattering indiscriminately and facing in our direction. Panza, assuring me of safety, kept walking ahead, but when a nearby cow lowered her head and slowly advanced with a stance much like that of a pointer dog approaching his game, I decided to make as nonchalant and casual a departure as possible. It was still too dark for photography, anyway. No animals followed us into the open.

At our closest we were probably thirty yards from the herd, and, according to Panza, their approach was motivated solely by curiosity and not belligerence. On all occasions, as I later found out, Panza's judgment of an animal's behavior proved infallible, but this was my first day with him and I attempted no psychological tests on these animals to determine their reactions.

After seeing the herd of buffaloes we set out in search of

rhinoceros, preferably the rare "white" ones. We found none, but during the morning alone we saw eight of the "black" rhinos, and I photographed a group of three at a distance of not more than thirty yards. This species had increased in number during the past thirty years, and so had the white rhinos, which had been induced to move over from their haunt in the nearby Umfolozi reserve.

I was able to approach close enough to the blacks so that at one moment it was impossible to get all three animals into the picture, even though they were almost touching haunch to head, at right angles to my camera. Eventually one of the three trotted toward us. Panza suggested a more or less leisurely retreat into a convenient tree. I accepted his suggestion with alacrity, especially as this animal continued to advance and came so close that only its forequarters showed in the camera view finder. I took one picture at a later paced twenty-two yards, not close enough because of the interfering brush, but close enough in terms of available trees.

The speedy and direct approach of this rhinoceros could scarcely be called a charge, and I scrambled up a tree only because I could not be sure whether his close investigation would not turn to belligerence. This new tameness, though not to be counted on by a stranger in the area, was a surprising contrast with experiences I had thirty years earlier. It was probably a result of the animal's freedom from molestation together with the frequent sight of human beings in close proximity. It seems to be a case of familiarity breeding tolerance, if not contempt. Nevertheless there has been no fundamental change in the underlying nature of these animals.

At that time of year the black rhinoceros were mating, and for two nights one pair elected to engage in their courtship just behind the rest huts. From the sounds it seemed apparent that one of the pair was slow in responding to the other, and there were periodic stampings, rushes through the tall grass, and then the sound of a double set of legs followed by pauses for steam-exhaust snorting and protesting grunts. The noises continued through the nights and quieted only in the early dawns. That it was a love affair was affirmed by the game guard, although he did not venture out into the night to verify the fact. Rhinos are notorious for attacking lights, and in the necessary darkness under which investigations would have to be made, Panza did not care to probe into the private affairs of these rhinos.

In the mornings we found behind the rest huts fresh rhino dung, which the black rhinoceros treat with a hind-leg scattering motion very similar to that of a dog scattering earth, while the white rhinoceros pile theirs in heaps and tend to use the same spot, so that old and fresh dung not scattered is the sign of the presence of the whites. Also, I have seen the blacks urinating by ejecting very forceful showers or jets, backward from between the hind legs to a distance of fifteen feet or more, and horizontally at about a two- or three-foot level; natives said that these were the males who are especially equipped for this startling feat.

After our first night in a rest hut we toured the reserve for several days. One day unusually good fortune revealed three white rhinoceros together in one close group, two adults and one nearly grown juvenile. It was too late in the evening for good photography, even on black and white film. Besides, there were three black rhinoceros, in plain sight, watching from a distance of two hundred yards. Although this was a long distance in comparison with other encounters, the same guard was anxious and explained that these were unfamiliar animals, and that one of them might be an "umpunyane," reputedly a smaller, more viciously-tempered variety. Since it was necessary to move in between the blacks and the group of whites in which we were most interested, half our time was spent looking over our shoulder to observe what the blacks were doing, and it was not reassuring to note that instead of leaving, they were approaching closer. As they closed in, even the game guard, the wise and experienced Panza, showed some nervousness, and after a few more minutes of looking alternately at the whites so close in front of us, and the blacks not far away, it was a relief to have him advise that it was time to leave. That had been my own feeling for some time.

Probably the most common animals in the reserve today are the wart hogs, *n'tibane*. They were present in singles, pairs, and family groups numbering as many as six half-grown piglets plus father and mother. The tusks of the adults protrude at a crossed-sickle angle.

Next in abundance to the wart hogs are probably the baboons, which in turn are far more evident, and probably fully as numerous as the more timid monkeys. We saw troops of a dozen baboons many times a day, roaming the open fields, rock ledges, and broken forest. They seem to rely more on their numbers and effectiveness on the ground than on escape to trees, but they do show a strong inclination to take refuge on cliffs ("krantzes") or on rocky promontories.

The baboon population is much too large. Their increase in numbers can be ascribed to the scarcity of leopards, which for many years have been poached by natives using steel traps. Leopards are of course protected in the reserve but so long as a leopard skin brings money amounting to a month's wages, poaching will continue.

Back of the rest huts the trumpeter hornbills (in South Africa called toucans or, by the Zulus, *amakunata*) returned each day, bleating or screaming their cracked-trumpet calls. I did not hear their full calls or songs until spring, nor those of the common turacos, plantain eaters, the *igwalagwala* or *igolomi*, which were feeding on the fruit of an *umkiwa* fig behind the rest huts. Here as almost everywhere in tree areas, the black-capped bulbul, or *ipotwe*, were common. Their ubiquity is another index of their versatility and adaptability. If introduced into California or Florida, they could become worse pests than the European starling or the Indian mynah bird.

PEST CONTROL AND ECOLOGY

Early in the morning of one of our days in Zululand, a number of small airplanes, at least six, were dusting or spraying for the tsetse fly in the nagana-control program. The planes take off from an air strip in a narrow valley and "buzz" the bush in the lateral valley bottoms. Since these strips of bush are sometimes very steeply sloped, narrow and tortuous in their turnings, and the planes must lay their dust strip low along the treetops, the flying is magnificent and demands a type of maneuver that is seldom called for in our American crop dusting.

An ambulance and crew were standing by, and it was a relief to have the program over with soon, without accidents. All dusting was completed soon after sunrise, and the released planes sped up our hill and cleared the ridge and rest huts by so narrow a margin that the prop wash could be felt.

No one seemed to know the chemical that was being applied in that particular operation, but hours later when we toured the sections that had been dusted I noted that the odor was similar to that of gammexane.

So effective is the destruction of the fly, that where the Harris trap at one time would yield as many as 1,500 flies a day, now five a month is a good take. This trap is barrel-shaped, painted black below with the sun shining through a screen from the top so that the flies enter through a slot in the shaded bottom, fly upward toward a seeming exit, and are thus caught. It is thought that ultimate extermination of the fly may be possible. However, in the meantime, it is not known what is happening to various species of other insects that may play an important, even a decisive, role in the total welfare of the reserve. Nor is it known what will happen to the insectivorous birds, the amphibians, the reptiles, and the useful parasitic insects.

Whatever undesired effects may result from this dusting or spraying, at least there is one minor compensation: The tightly sticking flies which, many years ago before the spraying program, crawled into ears, nose, and mouth, and clustered around one's eyes, seem to be no more. At least I noticed none, and even in summertime there is a surprising lack of troublesome insects. Thus, it is both safe and pleasant to visit the reserve at any time of the year. Even in the more salubrious climate of coastal Natal, pestiferous flies were more annoying than in this game area. How long it will be until resistant strains develop is an interesting question, but the authorities may alternate the types of toxins employed and thus make a total kill. The killing of the flies may be pleasant to visitors, but since the ultimate consequences are not predictable and may be disastrous, this dusting seems to be a needless risk of a possibly irreplaceable resource. It may be discovered too late that some fragile link, but absolutely essential in the chain of complex ecological interrelationships, may be irreparably broken.

Already from the viewpoint of a more or less casual though "old timer" spectator it seemed probable to me that the reserve was overstocked with some species of animals, and that the forage—grasses, herbs, and some types of small trees—was under severe pressure. Whether the summer growth could continue to overcome the winter deficit that resulted from overgrazing and browsing would be a matter of conjecture, and the problem of the welfare of game, forage, and even the soil a matter for careful ecological study.

To one who had been absent almost thirty years, there seemed to be notable changes in the abundance of some grasses; a few species of bushes undesirable as sources of food, appeared to be capturing ground that would have been richer in forage if these plants had been suppressed. One such species that seemed to be spreading as compared with earlier recollections, was a member of the Celastraceae, diagnosed by the local botanist as *Gymnosporia*, probably *senegalensis*. There seems to be no common name for it as yet.

There were few indications of any browsing on its glossy green foliage. In winter, with food so scarce and so many plants dormant and woody, the green leaves if they were edible, should have tempted all browsers. These plants could escape destruction only if in some respect they were either repellent or poisonous. It is said that the black rhinoceros will nibble on it occasionally, but if so there was little indication that such browsing was at all extensive or sufficient to prevent a spreading of an only slightly useful plant at the expense of better species. There were probably many other unperceived changes.

In the latter part of winter I found many animals in a noticeably emaciated state, and I scarcely doubted that this was due to overpopulation by these species themselves, or by the entire complex of competing feeders. Those suffer most which are least successful in a contest for food or which have a more restricted diet.

Preservation of most forms of big game has been so success-ful that overpopulation by wildlife in the reserves may become serious if it is not already so. Without data on the carrying capacity of any specific area before the advent of modern man and without knowing explicitly what plants were affected and to what extent, any appraisal of present conditions is impossible, and experimentation will be needed if optimum conditions are to be maintained. However, there is one notable and readily re-establishable alteration in the biota of the reserve—the native human population. These organisms originally were few in number but they were certainly an integral part of the "balance of nature," and they were exceedingly effective predators. By their employment of fire when hunting, or their use of it to provide early grazing for their cattle or as an attractant to the wild animals, they may have played a decisive role in the maintenance of "natural conditions." Their part in the "web of life" cannot have been a minor one in setting up and maintaining the balance between the savannah and bush, as well as the long chain of reactions between living things, that extend from soil bacteria to the elephants.

Apart from their widespread use of fire the natives hunted with spear and club, pit and snare, and naturally their activities were concentrated where wildlife was most abundant, as near watering places or good food. Man's presence must have been a disturbing factor that could keep the game circulating over the countryside thus relieving unduly prolonged or intense pressure on the food supplies of any one area.

The complete exclusion of primitive man from continued participation in the ancient pattern of life—the total ecology that had become established in eons of time—may at least in part explain the symptoms of unhealthiness in many reserves today. If man was important in the past, his total exclusion from the reserves today may well multiply troubles in the future. In any event, reserve ecologists will have to learn by trial and error how to substitute modern techniques in place of natural phenomena, a difficult accomplishment.

Although predator balance at one time might possibly have
supplied the required remedy, it is also possible that removal
of insect pests, including disease carriers, may be favoring this
unhealthy overcrowding—possibly mortality in the young antelopes of some species is too low.

The problems are complex, and it is fortunate that there is
now an ecologist appointed to study these reserves in great de-
tail. However, in fairness to ecologists, they should be allowed
years of detailed preliminary work before they are expected to
proffer satisfactory solutions to all problems raised by dearth of
predators, primitive man, and insect life. Most basic information
needed for interpretation of present conditions has been permanently lost for want of trained biologists throughout many
decades past. Conditions of at least twenty, even fifty, years ago,
would have to be known to detect and evaluate changes that
have already taken place, to make recommendations for treatment and rehabilitation of vegetation, and to appraise accurately
the normal or safe maximum carrying capacity of the reserve's
many species or organisms.

KRUGER PARK

During that winter I also made a short side trip to what is probably one of the world's best-known big-game sanctuaries, the
Kruger National Park in the eastern Transvaal, on the borders
of Portuguese East Africa.

The fame of this area is justified by the variety of the fauna,
the size, diversity, and accessibility of the terrain, the excellent
management, and the successful preservation of the carnivores
as well as the herbivorous game. Because of the unhealthy
summer climate, most of the park is closed during the malarial
season, but in winter it is open, and the nights are sometimes
bitterly cold—surprising for visitors who expect it to be tropical
and leave heavy bedding and coats at home.

From the standpoint of beautiful scenery the Hluhluwe reserve outdoes the Kruger Park and many other African big-game
sanctuaries and, until 1953, offered the incomparable asset of

providing a feeling of the wild—the privilege of walking through the bush and over the hills, unattended and unhampered, dependent on one's own good judgment as to one's welfare and that of the animals.

That South Africa could do as much as she has done, on so small a budget, and among a people who traditionally have been as great or even greater hunters than our frontiersmen in the United States, is a tribute to a few great minds, once called hopeless visionaries, and to their support by a large body of other sympathetic citizens of the country.

In terms of total area preserved in proportion to its population, South Africa should be credited with being in the forefront of game conservation, and South Africans deserve particular commendation for their retention and protection of the large and dangerous carnivores that are so essential to ecological balance in the reserves. This reflects great credit on the diplomacy of government in relation to the farmers of the country, especially those whose land adjoins the reserves; credit also goes to the naturalists, practical ecologists, and statesmen who first envisioned and worked toward the salvation of these areas.

It is exceedingly unfortunate that in Zululand the large predators even before my latest visit had already suffered such severe losses that lions and leopards were no longer effective controls in the Hluhluwe reserve; nor does it seem possible to introduce them now, over the protests of landholders whose livestock might be jeopardized. The danger to predators may develop elsewhere than in Zululand, because with pressure for more land by an ever increasing native and white population, more intensive farming will develop close to the reserve borders. If the nature of the farming changes to the growing of plant crops, pressure will develop for destruction of the herbivores as well as the carnivores.

In anticipation of these ultimate developments, provisions should be made for restricting certain potential farm areas to another type of gainful activity, namely "dude ranches," with hunting rights along the game-refuge boundaries. This is already developing to a degree. It will be a logical step so long as effective fencing is too expensive to construct and maintain, at least around the existing sanctuaries.

The "flavor" or aura of each of the African game reserves differs to a degree that is easily perceived by naturalists and observant laymen. These differences include as their bases not only the species of the big game that most tourists come to see, but in many little things such as the tribal differences and languages among the Africans resident or working in the native reserves. To an even greater degree the small animals, the bird "hangers-on," free bed-and-boarders, lend a different touch to individual camps as well as to the different areas. Some of these differences are obvious, others more subtle.

Just outside camp limits at Satara, in the Kruger Park, lions were lolling in the sun, indifferent to the many cars and chattering tourists who made the most of an opportunity to see these magnificent animals. Lions are admittedly more exciting than birds, but actually no more intrinsically interesting than the local avian camp beggars that were scattered over the camp grounds, almost completely ignored by the visitors.

In the United States jays are the most conspicuous camp frequenters, but in the Kruger Park their part is played by two species of glossy starlings, two doves, and the red-billed hornbill, frequently misnamed toucan. This latter bird is a local clown, always amusing and sympathy-provoking and vaguely suggestive of caricatures of our Western road runners. However, the hornbills are infinitely tamer than road runners, and they are more grotesque, with their short legs and huge, ungainly bills. They are also tamer than any species of their family and they will approach almost to one's feet to be fed with bread or fragments of meat.

On trips between camps in the Kruger Park, I also noticed that wart hogs, monkeys, and baboons were much less common than in the Hluhluwe, probably because of the presence of lions and leopards in the Kruger Park. Jackals were more frequent; these beautiful little animals strongly suggest our coyotes, whose ecological niche they almost duplicate in Africa.

Of the larger animals, the absence of both black and white rhinoceros in the Kruger and the presence of many giraffes, elephants, and lions in this northerly reserve, constitute some of the more obvious and notable differences between these two outstanding wildlife sanctuaries.

The regulation requiring visitors to stay in their cars is soon felt to be an irritating restriction to detailed observation of the animals that one has come so far to see, and even a naturalist, supposedly conditioned to think in terms of the welfare of the animals rather than his own interests, may at times feel that he is hampered in a perfectly legitimate desire for closer observation.

Few people seem to realize, however, that this rule enormously facilitates close and safe observation of all wildlife by encouraging it to linger beside the roads and near camp. Not only does it permit more intimate studies, but it also makes possible far better photographs in shorter time than would otherwise be possible. Furthermore, photographs taken from motor vehicles will usually portray the animals in more natural poses than could be obtained by the foot stalker. The reasons for this become clear: animals are by now so accustomed to motor vehicles that it is easier to approach them in cars than on foot, and they are less nervous when the passengers stay inside of their automobiles. Furthermore, in the bush proper, grass and twigs of brush generally conceal the animal's feet, legs, and often much more of the body. As the visitor attempts to maneuver around the obstacles, the animals become suspicious and alter their poses from resting or feeding to alarm posture. With many of them the reasonably close approach of a car has little influence on courtship and other rarely observed activities, whereas the presence of a pedestrian is usually detected and the animals abruptly terminate their natural behavior.

By far the best photographic records can be taken from a vantage point as high in a car as conditions permit, preferably from a porthole in the top of a truck. Most of the best photographs shown publicly for the past several years, as well as most privately shown films, have been obtained from specially made photographic cockpits placed in the tops of trucks that include chairs, rotating tripod mounts, or sandbags. With such conveniences, plus fast, long-focal-length lenses, animals a hundred yards away can be made to appear as though they were at arms' length.

From a high place one looks down into or behind most obstructing brush and grass and therefore has a far better chance of avoiding intervening foliage or twigs that might blur the photographs. The animals seem not to note the monkey-like object that moves a trifle on the back of the familiar but harmless apparition of a car. This tactic has procured by far the best series of lion, elephant, buffalo, and rhino still pictures or movies with safety to both man and beast. A car is seldom attacked but if it is one can usually escape a charge; on foot a charge often means a choice between death to the man or to the irritated or frightened beast.

Game of different kinds is often scattered over a wide area, and only by car can one cover great distances in a short time and thus discover the specific animals one wishes to observe. By contrast, when on foot game may seem very scarce as one laboriously toils through miles of brush and tall grass, most of it unoccupied by game. In a car these "vacuoles" are so easily traversed that they leave little impression, and one obtains a picture of animal abundance that the foot worker, with no car experience, might consider exaggerated or impossible.

For naturalist or casual traveler comparative intimacy with the animals is undoubtedly one of the great advantages of their study from a car, but it cannot wholly offset the very real value derived from a pervasive sense of danger that once was a necessary part of big-game study. Traveling by motor vehicle conveys practically no appreciation for the ways of primitive man when he lived surrounded at all times by these animals. This is a serious loss, although possibly not so much for biologists, as for the anthropologists, because it is only in this manner that they could truly appreciate the life of early man. No amount of reading about big-game hunts by white men can convey the different points of view that existed between well-armed white men and virtually unarmed natives. Thirty years ago, substitution of camera for gun did provide better understanding of primitive man's relation to his competitors.

These were the thoughts that came to me on my flying visit to Kruger.

5

BIRD LIFE AND THE SEASONS

Even where the big-game animals were long ago killed off in Natal, birds still flourish in seemingly countless variety and number. Ornithologically minded visitors wishing to observe under the least distracting circumstances would do well to spend a good part of their time in the remaining strips of bush and wild lands on farms and in the lesser sanctuaries. In the big-game sanctuaries they will be continually distracted by large animals, some of them dangerous, and certainly will be hampered by the now universal order not to leave their car.

Although a man may be birding in tame or tamed country where big game has disappeared, it might be advisable, or at least comforting, to carry a good stick as well as the normal equipment of field glasses, camera, notebook, and possibly tape recorder. The slow movements and the inconspicuousness of a good ornithologist might, especially in warm weather or just after a rain, enable him to creep so close to a basking mamba as to alarm it into aggressive action. However, despite this suggestion in all my collecting experiences and despite the number of mambas I have seen in the bush, I have never yet been attacked by this snake, except when it had been disturbed.

In all other respects birding in South Africa is much the same

82

as in the United States, even including an excellent textbook illustrated with color plates.*

For those familiar with our own Gulf states, I would estimate that numbers of species and the abundance of individual birds are somewhat similar which means that an American ornithologist visiting South Africa will encounter a large number of different species in any one area. If his stay is to be a brief one, the presence of unfamiliar families, as well as the welter of unfamiliar songs, will be confusing at the start of the visit. A visitor will find that the "sorting out" process that must precede a working familiarity with local species will be slower for him than it would be to a Northerner visiting the Gulf states, but even so working knowledge is rapidly achieved.

In the following comments on birds I have had no choice but to play favorites; there are too many species to mention except in a textbook manner, and I have therefore felt free to choose those that seem to me personally particularly charming, interesting, or, in some instances, scientifically worth noting.

As in all lands well north or south of the equator, in South Africa the change from winter to spring is often accompanied by violent storms, and these may erupt unexpectedly in the sunny calm of winter.

These storms in the midst of peace seem almost symbolic of South Africa, for it is possible that the natives may be triggered into widespread violence for no single explicable, therefore predictable, reason. The illusion of peace in the countryside can conceal potential dangers for the whites.

Through most of the year, this country is calm, and that seems to be its keynote; nevertheless, there is often wild, untempered fury of the winds. And this is not the localized violence of a tornado, but a country-wide storm with gales that blow for a day or more. The feeling of enormous forces, the overwhelming pressure and energy behind these widely moving masses of air make the storms impressive.

---

* Austin Roberts, *The Birds of South Africa*, H. F. and G. Witherby, Ltd., London, 1940. Obtainable through the Central News Agency, Ltd., Johannesburg.

But storms cannot last indefinitely, nor will the season between the droughts of winter and the summer rains keep up for long, and at last the welcome signs of regular rains appear and life prepares for the new season.

After the rains start, at long last the dark green of the bush that lies in indentations on hills and along the valleys begins to show faint pinks and reds against a shade of lighter green, and then the frog chorus swells to its diapason. The birds that have been unobtrusively scraping out a scant living by a great deal of appetite-engendering activity, begin to announce their presence with song and call. With increasing hours of daylight, endocrine changes set off a chain of events, and so when food is also more easily obtainable, there is desire, time, and energy for the extra effort of reproduction. Many secretive little birds that had been unobtrusive during winter, suddenly attract attention by their songs and courtship antics.

During these days of awakening life, the first evidence of spring appears as an influx of previously unnoticed birds. From overhead come the season's first wild, shrilling cries of big, black and white swifts, a heart-stirring sound. There are few birds whose notes are more thrilling than these high-pitched vibrant calls, uttered as the birds circle at tremendous speeds high in the air, sometimes almost out of sight. They suggest the whistling of violent gales through the rigging of a ship at sea, and connote the same sense of exhilarating freedom and joy at breaking the bonds of two-dimensional living.

By day and night, the birds increase in number and variety. In September in the South African prespring, when the clouds often cover the sky for the first time in months and the air is warm and humid, some early morning when the light is still gray, or some evening when the sun has set, one suddenly hears a song that to an American spells spring as surely as green grass and woods dotted with flowers. The song comes from a close relative of the American robin, the olive thrush (*umunswi*), and even the call and alarm notes sound robin-like. The thrush's "rain song" brings to mind pictures of cloudy days and dripping trees, when nest building in the United States is about to begin.

The song of the olive thrush is only reminiscent, however, for there is a little different quality to the music, a slightly reedy edge to the notes.

Spring in Africa as in any land is a season of joy and beauty, when all senses are stirred. The patter of rain on the roofs and the musical splashing of water running into the iron tanks at the corners of the houses are just a few of the joyous sounds. But these particular notes are of more than casual significance in this dry country. Drinking water is unsafe unless it comes directly from the rain and has not been contaminated by flowing over the ground, and when the dry season has been long, and the tanks have been emptied, all drinking water must be obtained from the rivers and then boiled. Water boiled, particularly on a wood fire, does not taste too good, and often it is only when the first rains have filled the tanks that it is once more possible to drink deeply and safely of a palatable water supply.

Not only do the rains symbolize such simple pleasures as good drinking water, they also mean happy natives. For their lives are dependent on good crops, and with a craving caused by having had no fresh food for months, the natives take rare joy in these moist days of spring and burst into spontaneous song as easily as the birds around them.

Although the winter is beautiful, and the sequence of cloudless days for month after month is delightful, spring brings more delicate beauties. At early dawn one looks down into valleys brimming with white mists, and through the cottony sheets of cloud rise the songs of the natives plowing in what is left of the rich valley soil. Men, women, and children share in the labor, and all seem to enjoy working with the mellow earth. They are in closer contact with nature than we, and the planting means a direct food supply, not a source of revenue. It means life-giving food itself, direct from the soil to the hungry mouths.

By October, spring has usually arrived. This is the time when the birds are becoming most vocal and the nectar feeders grow fat, and also the season for the fragrant perfumes of the growing bush. These are quite different scents from those of the winter with its spicy aromas, when the blends of plant oils are volatilized in the warm winter sun. By mid-October the bush bursts into bloom and the bell-shaped gardenia, lavender "syringa" (umbrella tree, *Melia azederach*), jasmine (*isandhla ka n'kosigasi*, hand of the white lady), and a host of other sweet perfumes flavor the air. A walk through the bush or a drive down the highway is an olfactory experience that calls for thanksgiving that we have retained this ancient heritage to a degree at least sufficient to enjoy this season to its fullest.

For a hundred miles along the coast the dark green *upahla* trees turn their backs to each spring breeze that blows, showing their silvery undersurfaces to the balmy wind.

As they follow the plow, the bare feet of men and women feel the soft texture of the mellow loam, and from the ground rise the odors of clean earth and promise of crops to come. It is no wonder that natives shout in joy, and those with cattle crack the long ox whips vociferously. Men and women work with smiles on their faces, the children take turns leading the slow, plodding oxen that drag the plows, turning over fold after fold of moist earth.

Between turns the children roll and tumble in the newly exposed soil, or follow the plow dropping seed into the ground and chasing off the doves and pigeons that settle in flocks to gather a share of the feast.

During the days of spring all living things seem to be courting and mating. In almost every patch of bush the birds are singing and dashing about in pursuit of each other. Some, the shy and retiring, advertise their presence only by song, but others perch on the tops of the tallest trees, on the bare branches that project into the open.

Even the pugnacious South Africa drongo shrikes, funereally black, temporarily show the stimulating effects of the mating season and spend hours dashing from tree to tree. They go in pairs, or quarrelsome trios, and sometimes four or five gather on the bare, topmost limbs of the trees and attempt to sing. They bob about and jerk in the agonies of producing song, and for all their prodigious strivings can emit only what is probably one of the world's worst caricatures of music—a clanging song like the sound of a squeaky and ramshackle iron wheelbarrow. Even the most sentimental listener would find it difficult to believe that such noise might win a mate. Only the fact that it is their specific song, their declaration of territory, and that the contortions are only part of the reproductive cycle could explain their success.

When hawks or other enemies alarm them, and only then, the drongos give a clear flute-like, melodious alarm call. They are most beautiful when in flight, for they are experts on the wing. Unlike true shrikes, these birds dart about in the air and, like flycatchers, hunt their food in flight. When they are wheeling about in the air or launching out from a tree in pursuit of some tempting insect, they are superbly graceful, but it is when grass fires are raging and they swoop and soar through the puffs of smoke, chasing grasshoppers and other insects that have been driven into the air by the flames, that they are at their best.

This habit of taking advantage of the insects being driven before the flames of a grass fire is not confined to the drongos, but they are the most ardent and also the boldest and most reckless of all the birds engaging in this practice.

After fires have swept the hillsides, and the fields are bare of cover, many birds gather to glean the harvest of food left by the scorching flames. The big ring-necked raven (*umgwababa*) is one of the most persistent gleaners, and on the black, ash-covered fields he blends perfectly with the background.

I once saw one of these big ravens winging back and forth over a recent and still-smoking burn looking for prey, and then swooping down into the black ash to rise a moment later with a snake four or five feet long. From its very slender form and black color, it appeared to be the dangerous black mamba or tree cobra. It was still alive, and as the raven headed toward the distant cliffs, the snake coiled and uncoiled in violent efforts to struggle free.

In the spring, a jasmine (*Jasminum multipartitum*) is one of the first bush flowers to bloom, and one of the longest to stay in flower. In the early mornings the air is sometimes redolent with its delightful perfume, and in the country where each fragrance blends with other subtle odors, it is at its best. Around a house there are too many alien odors for it to be fully appreciated, but in the fields even the barbaric little herder boys sometimes comment *"Ku ya nuka gahle ka kulu"* (It smells very nice). About the time that the jasmine begins to blossom, the *intsantsas* start their courtship. These inconspicuous little brown brush robins suddenly sing themselves into prominence. In almost every patch of brush they whistle their loud, clear song. To see them singing is as much a pleasure as to hear their song, for with each whistled note they utter, their rather long, white-tipped, fan-shaped tail is jerked a little more nearly vertical, and spread a little more fully. For an hour or so each morning, and then again late in the afternoon, they seem to sing most enthusiastically, but at any time of the day their song can be heard from some direction.

The vocal ability of tropical and subtropical birds has been grossly libeled. There is no lack of real beauty in many of their songs. Taken altogether, there is as much delightful bird music in Natal, on the edge of the tropics, as in the distant northern hemisphere. So much of what man regards as beautiful in a bird's song depends on his associations and sentiments, that it is not surprising that the birds of tropical countries, any foreign country for that matter, have so often been maligned in regard to their song. One of the most common sayings—"South Africa is a land of rivers without water, birds without song, flowers without perfume, and women without beauty"—is very far astray in every instance, and nowhere more so than in its allusion to the bird songs.

SUMMER

With the passing of spring the days grow longer and the sun beats down more fiercely, thunder storms and rainy periods increase in frequency, and the grass grows tall and green. Throughout the bush a ground cover of weeds, most commonly of one species, *mabomane* (*Isoglossa woodii*), becomes rank.

In most parts of the world summer brings tormenting gnats

and mosquitoes, but here they are rare. Even in the temperate zones of the northern hemisphere—the Sierra Nevada of California or even Alaska and Canada where the winter cold is intense—summer brings myriads of biting insect pests, "punkies," mosquitoes, and horseflies, while house flies and other obnoxious pests among the nonstinging or biting varieties multiply rapidly. It is surprising that over so great a part of Natal, close to the malaria-ridden areas not far to the northeast, there should be such an extraordinary lack of disagreeable insects. Usually only the house fly occurs in sufficient numbers to be a nuisance and a threat to health.

Since the dipping of cattle has become a universal custom, even the ticks, a previously ubiquitous pest, are not so noticeable as they once were, and one can walk for miles through grass and bush with less inconvenience from insects and their allies than in the temperate zone of the United States. There are no "chiggers," and a day in the grass and bush of Natal is far less discomfiting than in Washington, D.C. or the Southern states.

As the days become hotter, the air steams with moisture. The early mornings are just cool enough to be comfortable, but warm enough to bring out the intoxicating odor of damp earth and wet humus. As one walks along the edge of the bush the damp air is heavy with many delightful scents. Most of the fragrance is neither sharply pungent nor cloyingly sweet, merely suggestive of luxuriant growth and clean country soil. Here and there a vagrant breeze brings the catnip-like odor of the crushed leaves of a verbena, the *umsuzane* brush.

Another scent is the arnica-like, pungent smell of a species of termite that builds its wood-carton nest only a few inches above the ground, and so perfectly matched to the soil as to be almost undetectable. This scent is unique in its suggestion of antiseptics and the sharp but pleasant smells of a drugstore. These termites are the most sluggish representatives of a sluggish order. Their termitaria are rarely a foot in diameter. The insects swarm through their galleries in large numbers, and some of the cells are packed with the very slowly moving white bodies. Unlike other species of termites, these have no effective soldier castes to act as guardians for the soft and helpless workers, and if a nest is broken open, the termites slowly make their way into the deeper recesses where they cluster together for hours or days, until they fall prey to the little red ants that teem over the floor of the bush.

This inability to travel freely, combined with the habit of taking refuge in the fragments of their broken nests, are traits used to advantage by the native boys. Termites in general, living such sheltered lives, seem only rarely to have adequate methods of self-defense once they are removed from their hiding places. Unlike so many other seemingly helpless insects, they do not appear to have any repulsive taste, nor does their "drugstore" odor seem to be repugnant; and they are evidently not poisonous to any of their varied enemies. Without their habit of building clay or wood-carton shelters, their succulence would long ago have led to their extinction. One and all, they appear to be favored as food by almost every carnivorous or insectivorous animal.

The boys in Natal set their bird traps in the bush where there is sufficient humidity and shade to prevent the termites from drying up. They take advantage of the helpless condition of the termites by breaking a nest open, tapping two nest fragments together, and shaking the inhabitants over the ground near the trap. These termites, when removed from the safe chambers of their nest, remain for several days creeping slowly over and under the leaves in the immediate vicinity until they are eaten by birds and other animals.

The termites make excellent bait for birds and attract almost all bush-dwelling species accustomed to feeding near the ground. The caught birds are eaten by the young natives.

Doubtless other termites would be equally attractive, but all other species are too active to be useful for bait, and disappear from the ground where they have been sprinkled. The one species that does not crawl away is so immobile once it is removed from the colony, that even though fragments of its nest are less than a foot away from the area where the termites have been placed, they never seem to discover the fact or reënter the shelter so close by.

In addition to thrushes, babblers, and the bush-dwelling doves, the Cape bristle-necked bulbuls or *bwadas* are among the birds most frequently attracted by the termites. The birds travel in small bands of three or four, sometimes more. They often annoy hunters, because in the thick bush it is impossible to see very far, and when watching a game trail, a hunter is often startled into excited expectation of the arrival of an antelope by the rustle of leaves. At first the loudness leads him to expect at least a bushbuck, but the very fact that the rustling is so loud should indicate that the maker of the noise is small. Unless antelopes or other big game are panicked, they seldom make much noise when moving about. In addition to scratching on the ground and rattling leaves, the *bwada* makes its presence known by one of the most distinctive calls or songs to be heard in the bush, a series of low, throaty clucks. If one can imagine a meadow frog warbling, one can gather an impression of this song.

The Natal robin chats, and the wonderful noisy chats (both called *ugaga* by the natives) are other frequent victims of the fatal attraction of termites used in the traps, and both are among the many beautifully colored birds inhabiting the bush where traps are commonly set. They are also among the best songsters. Ordinarily the noisy chat makes just another beautiful song, but at times it mimics other bush birds, and on such occasions it pours out one of the most rapturous of all bird medleys. Five or six species may be mimicked in one outpouring, sometimes in a continuous song, but more often interspersed with short intervals between each variation of the performance. Most startling of all is the frequent finale to the song, a series of single, imitative notes. In this finale, there may be the plaintive peeping of a little downy chicken (this I have heard only near chicken runs), the sharp notes given in the first part of the emerald cuckoo's song, the calls of the red-winged starling (*intsomi*), almost the entire song of the black-headed bush shrike (*umquapane*) given in fragments, the calls of the black-headed oriole (*umqoqongo*), and various other songs and calls.

On one occasion, as I whistled for my boy, there was a low whistle in response. It came from a very thick patch of bush close by and was so man-like in quality, so low and secretive in tone, that it seemed as though it must have come from a human source. Under the circumstances this could not be, and so I repeated the whistle and again came the response. For several minutes this continued, and each time I moved closer to the source of the sound, always getting an immediate reply—the same throaty whistle. When I finally saw the mimic, it proved to be a noisy chat. As the bird answered each whistled call, it would spread and slowly depress the tail. That a bird with such richly colored plumage, rusty red and slate gray, and one so superior in singing ability, should be added to the gastronomic experiences of a little black savage is one of the tragic ironies of the bush.

As summer advances, the birds seem to increase in number. Evidently there is a steady influx from the north, a regular migration. Some of the birds do not nest in Natal; apparently they are visitants rather than residents, but there are many species that migrate elsewhere during the winter months and return in the spring and summer to nest. The migrations are not very well defined when compared with the more obvious habits of North American birds, but one notices the increasing numbers of birds and nests in the summer.

One early morning as the mimosas blossomed and the air was sour-sweet with their perfume, I was awakened suddenly by the sound of a new bird note, and on tracing the sound found a beautiful little paradise flycatcher, *uve,* darting about in the shade of a heavily foliaged tree.

The males of these little sprites flaunt an extravagantly long tail, and as they flit around in the bush or the deep shade of the more open spots, they appear to be flashes of dull-red flames.

It must be difficult to keep such a long tail looking so immaculate at all times, but like our own cedar waxwings, this species seems to warrant the description of "being always well groomed." It would seem as though long tails must be a handicap to such active birds, especially when they live in the bush and catch their prey on the wing, but so far as I could see, the male thrives quite as well as the female which has only a normal-sized appendage.

When I heard the song for the first time—a pert "Swei? Swei?

Swei-aaah"—I looked for a larger bird. I never had any difficulty in seeing the makers of the sound. Wherever they move, they show up. Even in dense shade, their bright chestnut coloring and sometimes even the bright blue eye wattle, are conspicuous. When sitting on her nest, the female is plainly visible. One so fortunate as to find her beautifully made nest may hope for successful photography, but almost always it is soon apparent that all interested animals are equally successful in finding the nest. It is surprising that any species of bird can survive such extensive nest predation.

The nest built in the fork of a twig some five feet from the ground is shaped like a wineglass, attached by the narrow waist to the twigs supporting it. This type of nest, built of dead grass and rootlets, looks exactly like a little clump of fine driftwood floated into place by a flood. It is probably by accident rather than design that these birds frequently build their nests in trees growing on low ground that is subject to flooding. Such nests are adequately concealed by the camouflage of resemblance to other multitudes of similar clumps. It may be that there is a higher survival rate among these than among birds that nest elsewhere, but there are no data to show existence of any trend toward favoring low ground.

Usually when I attempted to obtain photographs of birds on their nest, the construction of a blind nearby was tantamount to condemnation of that particular nest. Undoubtedly the blind attracted the curiosity of those ever-present marauders, the little native herder boys, and where these could be ruled out there were always the even more destructive monkeys. I built scores of blinds on my trips, but practically all of about fifty nests of various species of birds were destroyed by these two culprits before I could take pictures. Only a few nests remained intact, merely losing the eggs, and in these instances presumably snakes were responsible. In a land where the snake *Dasypeltis,* a specialist in the art of egg eating—in fact, dependent on eggs for food—is fairly common, the suspicion was probably well founded, even though I never caught one in the act.

During the hot summer days when the air feels lifeless and

dull, the first of the large migratory cuckoos are apt to arrive, the *nkanku*, as the Zulus call them. At ten or eleven o'clock in the morning, with no warning, out of the sweltering bush comes the doleful call, "paw, paw, jam," slowly repeated with a rising inflection. Many cuckoos are not particularly beautiful except in a quiet harmony of dull tints, but the interest they ordinarily lack in color is compensated for by their unusual habits. As a family they seem to have developed the parasitic trait to an unusually high degree, even though many species seem able to rear their own broods and do so successfully. In a broad sense the group ranges in habits and appearance from the road runner of the western United States, a long-legged clown marvelously adapted to the life it leads, to the small, gorgeously colored metallic cuckoos of Africa. These are dainty, brilliant creatures with clear, whistling notes, utterly unlike the usual cuckoo calls so well exemplified by the tonal quality of the cuckoo clock.

One of these metallic cuckoos, the southern emerald cuckoo, figures romantically in the native folklore. It is supposed to be thoroughly aware of the many drawbacks to matrimony, and it is said to be fond of flying into the trees above some love-making young couple, where it whistles to them, *"Umtwanyana, ungendi; umtwanyana, ungendi"* (Little child, don't marry). The origin of this belief probably lies in an onomatopoetic interpretation of the call notes, but the natives may also be familiar with the habits of the bird, one of the parasitic species. Even a white man familiar with Zulu, readily sees the similarity between this song and the Zulu words. Once I heard one of these little cuckoos in a tree overhanging a mission church during a wedding ceremony, and for half an hour it repeated its exhortation. Probably only a few, the back-sliders or heathen, fully appreciated the humor of the situation.

When the summer rains pour down heavily day after day in lengthy periods, all wildlife seems to suffer. Even the rain coucals, which commonly appear in ordinary rains still seeming well groomed, gradually begin to lose their sleekness and show by their bedraggled tails and wet feathers that they are losing the battle with the elements. A few of them become so chilled and waterlogged that they are almost incapable of escaping if persistently chased; some have been reported as caught by hand.

These rain coucals are often heard calling during the morning and evening hours. The notes are low, mellow and flute-like, trilled down, and, sometimes immediately afterward, up the scale. Their song has a dovelike, melancholy quality, is widely heard throughout the land, and is apt to be remembered as well as that of our own ground cuckoo, the road runner of the southwestern United States. The coucals are important for their place in nature as consumers of insects, reptiles, and amphibians, and probably also of rodents and young birds.

Native men and boys, who eat practically any form of meat, absolutely refuse to eat coucals for fear this may alter their sex. Women do not appear to have this fear, but for some other reason which I could not determine also will not eat them. On one occasion a group of men was horrified by my eating one of the birds, thus defying the superstition. For some time they stood about and watched, then with clicks of disgust said that it must be that the white man was immune to the evil powers or possesses a powerful charm. This is the universal answer when a white man demonstrates the falseness of some native belief.

Bird Watch by the River

After every period of rain when the sun comes out once more and the valleys steam with filaments of thin mist, the birds break into song and appear in far greater numbers than one would believe existed in the neighborhood. On every side they can be heard singing or seen dashing about through the air in pursuit of insects that rise in multitudes after each rain. At such times one of the most interesting places for my nature studies was some riverbank sufficiently open to walk about and watch the wildlife. Animals, like humans, seemed to be filled with rapture when the sun came out. They were then bolder and easier to observe than at any other time. Along the rivers, birds as well as all other forms of life, would congregate in numbers, and for sheer abundance of opportunities to observe what was taking place, these were the best localities.

In the rushes and reeds along the riverbanks, I would see weaverbirds working on their nests in little colonies. The nests were varied in shape according to the species building them. One of the most common was a beautiful affair made from the blades of the reeds themselves. The leaves or blades are stripped, a few fibres at a time, and are then carefully woven together until a nest has been constructed so fine and compactly built that only the roughest handling will break it. When the material is first put in place it is green, but a few days of sunlight soon turn it to an even brown, and from the time the nests are dry until they are deserted, they are among the most conspicuous objects along the reed beds. It is surprising that nests so plainly exposed are left unmolested, but except for the depredations of the native boys, they seem fairly safe. Doubtless the birds have some enemies, but these probably do not hunt altogether by sight or such conspicuous objects would soon be destroyed.

One species of weaverbird attaches its nest to the tip of the long fronds of the wild date palms. This palm has a formidable barrier of thorns along the base of each frond for a distance of one or two feet from the trunk, and the thorns from neighboring fronds crisscross and project in divergent directions, so that altogether it seems impossible that any monkey could negotiate them without considerable discomfort. If this mode of nest building really keeps out monkeys, the location chosen by this species should be one of the safest places. The spectacled weaver (*Ploceus ocularius*) mostly builds nests that hang over water, and they are often attached to the end of palm leaves.

The trees above the rivers were often alive with birds I liked to watch. The black-headed weaverbirds would be busy pulling off fragments of bark to pick up hiding beetles. Warblers and flycatchers would explore the leaves and twigs or snap up flying insects. Wood hoopoes commonly known by the Afrikaans name of "kakelaars" would glean what beetles and other forms of life were available on trunk and main branches. No niche was safe from a wide variety of specialized hungry animals; from the vegetation upward through a succession of small to large predators, there was a continuous transformation of protein, a transmutation from one to another species, and daily, even hourly, continuous reincarnation.

These kakelaars, or cackelers, are amusing birds and at the same time beautiful. They suggest in some respects small magpies, for they are iridescent bluish-green and black with white on the wings and long tail, and like magpies keep up a noisy chatter as they move about in small parties. The natives call them "laughter of the old women" or in their language *hlekabafazi*. This name is suggested by the shrill babbling which is of course their song. The birds ordinarily pair off when singing and bob and bow at a tremendous rate. The long, red, scimitar-shaped bills are held partly open while the birds go through their curious performance. As the notes come faster and rise higher in pitch, the bobbing becomes more violent, until suddenly at the climax they stop like an unwound clockwork toy. Thereafter they casually resume their hunt and move off.

When caught, the birds have an extremely unpleasant odor. At first one is inclined to believe that the specimens have unaccountably become decomposed, until one learns that this vile stench is connected with the oil gland at the base of the tail. In fact it seems as though their uropygeal gland might have the unusual function of secreting a protective odor.

In the river itself, I often watched the hunt proceeding as remorselessly as on land. The birds would participate in harrying prey through the clear water, and along the pools the beautiful and tiny midget kingfishers would sport their gorgeous blues and coral reds. From higher perches, giant kingfishers would scan the water and now and then go crashing into it after their prey. I once observed a pair of these large kingfishers habitually feeding in a small eddy of a river where the water was shallow. When I caught them, I found that one had a damaged eye and its bill was badly worn, as though from striking against the sandy bottom. The other with normal vision and feeding in the same shallows had a normal bill. It seems probable that the worn bill was not purely coincidental, but that it was a symptom of loss of binocular vision and depth perception.

The fishes were not only harried from above, but also under

the surface of the water by the interesting diving anhingas or snake birds, called *ikeus* by the Zulus—an onomatopoetic rendering of these birds' calls. They and the small reed cormorants form the most abundant ranks of underwater fishermen.

One summer day I watched from the vantage point of a high cliff, looking down into clear water and following every movement taking place in the pool below. It was surprising but not strange that, although I could plainly see the long, slender bills while the anhingas were under water, I could not see the fish on which they preyed and which are longer than the bill—a good example of protective coloration. Both from above and below the fishes were almost invisible, and their dull, grayish-green backs and bright, shiny lower surfaces seemed perfectly adapted to the environment.

While hunting under water, the anhingas cruised slowly about with wings folded against their sides. It may be that at times they use their wings for propelling them at a higher rate of speed, but during the periods they spent below the surface I saw them use only their feet for propulsion.

I observed one of the birds for a half hour, and it was interesting to note how easily so large a bird hunted for prey in the circumscribed surroundings of a boulder-strewn pool. As it moved slowly around under water, the long neck was partly outstretched, and head and neck were turned deliberately from side to side as though searching for food. From above, the bird looked much like some long-necked turtle and it seemed to be just as much at home. After searching out the deeper and more open parts of the pool, it turned toward the bank and swam about exploring the undercut sides for prey. At times it disappeared from sight, slipping under the overhanging banks, and then reappeared two or three feet away. It must have been investigating every possible hiding place. Once more out in the open stream, it managed to catch a little fish, and then rose to the surface holding it in its bill. For a minute or two the bird played with the fish, tossing it into the air and catching it again. On the last toss, however, the fish fell into the stream and got away. The anhinga seemed not at all disconcerted by the loss, and for a minute or two played about in the same spot, then dived, followed the banks of the stream down-current, and came to the outlet of the main pool. With no hesitation and still under water, it darted down some rapids with a pause now and then to examine some possible hiding place for food. The bird preferred going through this narrow, shallow stretch of fast water while submerged, whereas most swimming birds either would have remained on top or flown to the next pool.   */ 0 4 5 9 2*

When disturbed, the anhingas are able to submerge most of their bodies and keep only the head and neck above water. At such times, their vernacular name, "snake bird," is particularly appropriate. When not alarmed, they swim with the body floating high on the surface like many other water birds. When near the banks or in a dangerous place, they keep at least partially submerged. In contrast the little reed cormorants do not seem to submerge part of their bodies even when alarmed and seem to prefer flight to diving for safety. Judging by observations, the anhingas are quite as likely to escape under water as by flight.

The Crowned Hornbill

Only a few of the more interesting birds of the bush have been mentioned, yet there are many others that for one or more reasons should be included. Because of their unique nesting habits the hornbills as a group should have at least one representative here. Of the various species resident in the bush country, including the bass-voiced ground hornbill and the screaming-voiced trumpeter hornbill, none is more familiar nor for most of the year a tamer species than the crowned hornbill, *umkolwane.* Unlike most of the family, members of this species turned up regularly each year in the back yard of the mission where for days or weeks they flopped clownishly about the trees or fed on the ground, where they would wipe out entire colonies of the horribly smelling, brilliantly colored, flightless grasshoppers that not even domestic fowls will touch.

During the breeding season in the summer, the small flocks or family groups of four to seven break up, and the individual pairs are widely scattered and hard to find. Locating the nests of these birds is not only a baffling experience, but it usually involves a great deal of arduous work, normally during steaming hot weather.

In summer, the sun beats down unmercifully into the bush-clad valleys where these birds nest. During the warmer hours of the day most of the trees look parched and shriveled, especially the species of thorn trees that fold their feathery leaves and so obtain protection from severe insolation and evaporation. Especially during noon hours, the open glades are often dazzling in their brightness, and the air quivers with heat waves. On clear days between rains, the air is usually humid and oppressive, and as common in such weather, the only sounds are the occasional distant growlings from the massive thunder clouds and the persistent gasping whistles of the bulbuls. Even the rivers in the baked brown plain look hot and somnolent, irregular streaks of burnished silver set between banks of dark, flat-crowned thorn trees. Sometimes the tinker barbet, a tiny yellow and green sprite with a scarlet crown, may chime in with an hour of pulsating notes (fifty or so per minute) that throb rhythmically like the beating of a muffled metal bar.

But I was not after the bulbul or the tinker barbet. I had set my mind on finding an *umkolwane* nest, but it took several days before I succeeded. I discovered it because, in spite of the heat, one active crowned hornbill repeatedly flew to a certain section of the bush. This regularity ultimately led me to the hollow tree where the nest was.

The nesting habit is unique to this group of birds. When a satisfactory hollow tree has been found, the female begins egg-laying, and for a few days behaves in normal fashion. About the time the last egg is laid, however, the female retires to the nest, and with the help of the male, walls up the opening until only a small slit has been left, through which her mate is obliged to feed her.

Some people thought that the male alone carried on this plastering activity, but by sawing through the layers of cementing materials composed partly at least of fecal matter, I could see from the nature of the cement that the female helped to some extent. The walling-up makes her incubation a period of almost perfect security, for the nest is safe from all enemies—she successfully sacrifices mobility and freedom for this safety. Even snakes large enough to injure the incubating bird are effectively prevented from reaching her, and those small enough to crawl through the opening would be too weak to survive her defense or possibly her hunger, for she is well armed with a large and powerful bill.

During this voluntary incarceration, the female molts all or nearly all her feathers, and for a short time is almost as naked and helpless as the young. This habit of molting probably places a considerably greater than normal burden on the male, for in addition to supplying ordinary needs, it can be surmised that more than normal amounts of food are required to sustain the molting and incubating parent although she is more or less immobilized.

The female remains in the nest until the young are able to feed themselves, and then, according to reports, is able to leave the nest, breaking out with the help of the male. The parents then re-cement the opening to extend the period of security for the young. Both parents for a time feed the fledglings, then open the nest and permit the family to leave. There is supposed to be a final cementing, this time presumably to protect the nest from preëmption by bees. The same hollow tree is used for several successive years, and unless the opening was cemented shut, bees would probably take over the site.

Because of this closure of the nest-opening and the extreme wariness of the male during the breeding season, it is very difficult to locate the breeding sites. The entrance of the specimen I observed was so carefully plastered that I had to follow my nose as well as my eyes to establish definitely its place and identity.

Shinnying up and hugging the tree trunk, I noticed a strong birdy odor. This, followed by a minute inspection, revealed the narrow aperture and cement.

At all times throughout the breeding period the male was so wary and retiring that it did not seem possible he could be one of the same breed that had been frequenting the trees in our garden through the rest of the year. He was so shy that it was necessary to make extraordinary efforts to hide the camera for taking his picture. Even the observer's blind, from which I extended strings to the camera, was carefully concealed some distance away from the nest.

Throughout the succeeding days of observation, the bird was at all times very cautious and, when at the nest, exasperatingly panicky. Time and again he would come with food, only to spend an hour or so examining every possible site of an ambush with sharp, suspicious eyes. At first he would hold the food in the tip of his long bill and merely sit and watch and listen. He then would begin to fidget and move from tree to tree, and at last, with a toss of his bill, would many times swallow the food he had so carefully brought for his mate. For a moment after swallowing he would look about, then wipe his bill and look about again. His movements when wiping his bill were quick and nervous, and the large, red beak, when hastily rubbed on the tree branch, sounded like a celluloid soap dish rubbed on a comb.

When he finally generated sufficient courage or desperation to approach the nest with food, he did so only by nervous starts and jerks; a final hasty hop, and he was clinging to the tree beside the nest opening and passing in the food. At one such moment I clicked the camera, and with the click came a fit of hysterics. During all the hours of observation he had been silent, never calling when near the tree, but the sudden sound so close was too much. He gave a loud shriek and flopped away across the glade to sit and "shout" and "scold." Since no enemy appeared, he at length slipped away again, as silent as a shadow.

At first his timidity and watchful waiting were a constant exasperation, but with a little analysis of the circumstances, it became obvious that, were it not for this unusual care, there would be no crowned hornbills left. Unlike most other birds who soon find another mate when their own has been killed, the female hornbill would be totally incapable of obtaining another helper to save her; helpless and half-naked and sealed within the hollow tree, she would simply die within her cell, and, with her, all the young. The death of the male would inevitably insure death to all other members of his family. It has only been by extreme caution near the nest that a bird which is ordinarily tame has managed to escape extinction.

Many times while I watched the hornbills, the monotony of waiting in the blind would be broken by a steady rustling, the swish of a released branch, and a sound of dead twigs falling—advance notice that a troop of monkeys was approaching. Usually these were the only sounds to announce their coming, sometimes not even these, but simply the thud of a hard-shelled *n'gola* fruit as it struck the ground. The foraging band would come stealing along toward the blind, and soon bright black eyes in soot black faces would be staring intently at the object of their suspicion. If I kept still, the advance would continue until all the troop would be feeding in the trees about the hornbill's nest. There was never any sign of fear on the part of the hornbills, and no sign of interest on the part of the monkeys, which seems to prove that both have learned mutual tolerance. Monkeys are the greatest thieves in the bush, and they love young fledglings with a craving that suggests almost continuous protein hunger.

Often I would hear other rustlings following close behind the noises of the monkey troop, and if I waited patiently in my hornbill blind sooner or later a dainty little bush duiker, *ipiti* (an antelope), would mince into view. They are slate blue, about thirteen inches high, and bear a pair of miniature but perfectly formed horns. Their minute track is just about the size of the end of one's little finger. Except for their hoofs and horns and their dainty and neat shape, they do not seem at all like antelopes. They can be carried with little trouble in the rear pocket of one's hunting coat. These beautiful little animals seem to have a well developed affinity for association with monkeys, the worst hooligans in the bush. It is undoubtedly not affection that induces this association, but it may be that the antelopes rely on the monkeys for giving timely warnings of concealed or approaching enemies. From their high vantage points the monkeys have a far better view of all surroundings, and of all the mammals only the primates seem capable of understanding or interpreting the nature of a motionless object such as a leopard lying in wait. Even if no other advantage were derived from the close association, the monkeys leave a hot scent to confuse trailing enemies. Also, there is food, usually berries, to be gleaned from the ground under the feeding monkeys.

The opportunity of watching the interesting monkeys and duikers was often a welcome distraction during many long hours of lying in wait for the comings and goings of the hornbill.

Parasitism

Among the most fascinating birds I watched in Natal are the honey guides—the lesser honey guide because of his amusing egg-laying performances and the greater honey guide for his once legendary relation to man and bees.

The lesser honey guide is an undistinguished little bird, lacking the widely publicized habits of its larger relative. The lesser bird seems to have little to commend it to nonornithologists, except for its parasitic habits which make it well worth occasional watching for a few hours during the breeding season.

Parasitism in birds, the insinuating of the eggs of one species into the nest of another, the host species, is best known in the United States by the cowbird and in Europe by the cuckoo, but in Africa there are many species that have adopted this manner of shirking arduous family duties. The seeming resentment of the host birds and the energy with which they attempt to counter the efforts of the parasite so closely parallel what we would expect in our own species that it is difficult to portray the combined activities of these conflicting species without adopting terms that we would apply in our own case, thereby inadvertently presenting a misleading picture of motives and feelings that we might impute to these animals.

The behavior of a bird defending its nest against the attempts of a would-be parasite inevitably demands the use of the word resentment. The irritability, excitement, violent pursuit and attack, and the literally breathless pause between forays against the interloper, all seem to denote hate and anger, and yet it is inaccurate to ascribe these feelings to a bird whose actions are probably better described as "an evolved adaptational negative reflex to either invasion of the nest site or to oviposition" —a most unpicturesque way of quickly summarizing and evoking a mental picture of a most interesting phenomenon.

Among the more amusing episodes observed in the relationships of parasite and host are those of the lesser honey guide and its common host, the more robust, hole-nesting collared barbet or *i'smakele*. This barbet is about the size of a grosbeak, has a large, somewhat toothed or serrated bill, a red throat and face, and a black collar. The rest of the body is gray, brown, and olive green, in pastel shades, plus some additional markings. Its common parasite, the lesser honey guide, is a nondescript little bird, quiet and elusive at all times except when attempting to lay its eggs among those of its red-faced host.

Before egg-laying, the parasite loiters near a barbet's residence, a hole in a tree trunk, and makes inconspicuous reconnoitering forays toward the nest site, scanning the neighborhood in obvious fear of the host.

From their behavior it appears as though the barbets are fully aware that the honey guide's presence so near their nest constitutes a threat (actually they are simply reacting with hostile gestures toward the parasitic bird), and they repeatedly attack and attempt to drive the irritating object from the vicinity of their nest.

From time to time when all seems safe, the intruder, as if to make sure that the female barbet has not left unobserved, approaches the entrance and peers cautiously down into the nest hole. Almost at once, from somewhere in the trees overhead, one or the other of the watchful legitimate owners dashes angrily down and swoops at the offending honey guide. One of the barbets is usually content to pursue it only for a short distance, but sometimes an individual from within the nest, possibly the female, is not so easily put off, and each time she pops out of the hole, she dashes furiously after the fleeing interloper. These ordinarily phlegmatic species reveal unexpected dexterity, flying up and down, high and low, and round and round, twisting and turning in their efforts to catch or elude each other, but the honey guide seemingly flies with greater ease and appears to be making only enough effort to keep barely a short distance ahead of her pursuer.

When the barbet eventually gives up, usually not more than a hundred yards from the nest, the two birds race to reach the nest hole first. The barbet usually wins, possibly because the intruder dares not enter while the rightful owner is close behind, but the intention is obvious, and both birds take the race with desperate intensity. After reaching the entrance and taking possession, the female barbet pushes her head out of the opening, and with beady black eyes glares about in search of the offending nuisance. At these times, panting with her beak open, and with the scarlet neck pulsating breathlessly, the barbet suggests an avian caricature of an infuriated little old lady with high blood pressure.

During the whole proceeding, both birds are so thoroughly engrossed in their own interests that they can be watched from a distance of only five or six feet. The honey guide pays attention to nothing but the barbet. Probably an egg is fully ready for laying and is being retained only with difficulty. The barbet is equally determined to prevent the honey guide from laying. I never had occasion to observe the end of this contest.

It is a popular fallacy to denounce parasitism in birds, or any other organism, and it is usually done in order to point a moral illustrating how dangerous parasitism is to the individual practicing it. This viewpoint is scarcely in line with the facts.

Parasites seem to do rather well, in fact almost as well as any other organism, even those with the most orthodox methods of existence. Morally, in spite of human indignation, parasitism as a habit in birds, other animals, or plants, is no more reprehensible than the predatory behavior of lions or of hawks and owls. It is simply another natural way for an animal to perpetuate itself and of course does not imply laziness, carelessness, or any moral attribute.

The widespread human dislike of the cowbirds for their habits, or of the cuckoo, or honey guide, is possibly a thoughtless reflex resulting from human fear of members of our own society that become parasitic. The carrying over of this emotion into the realms of ornithology, where one species may parasitize another seems entirely unreasonable.

Parasitism is a common trait, and practically no animal lives without some other organism finding a living place within or on it. Parasitism is one of the most fascinating fields of natural history; there are problems for the taxonomist, ecologist, evolutionist, epidemiologist, and others, but probably in no other field of parasitology is there such appealing material and aesthetic reward as can be found in this habit as practiced among the birds.

Parasitism in South Africa can be seen on all sides and in all degrees, and it is clearly not an exclusively invertebrate or avian habit. The white mistress giving over almost the full care of her young to a native girl, and the Asiatic with his concubines and resultant children whom he puts to work so as to add their earnings to his, are all symptoms of a widespread tendency to foist off personal duties on others. Parasitism of this type is international, but among the birds it is more familiar in Africa than in the United States. In Africa parasitism is practiced by a finch, the pin-tailed whidah, as well as numerous species of cuckoos, among them the gorgeous little emerald *umtwanyana* and two other metallic species, and also the *pietmeinfrou* cuckoo and its kin. There is also the lesser larceny of the barn owls stealing the nest of the hammerhead stork, and the swifts preëmpting the nests of swallows. It is therefore not surprising that the European cuckoo, whose ways are probably better known than those of any other avian parasite, excluding possibly the American cowbird, finds the South African "climate" to be a salubrious one in which to escape the northern winter.

The Greater Honey Guide

A relative of the parasitic lesser honey guide is the greater honey guide, famous in fiction and scientific reports for its alleged habit of leading men to a beehive, where it is said to wait its turn for wax and bee grubs, on both of which it feeds whenever its efforts have led men to rob a hive. There has been some skepticism whether the bird lives up to its reputation. This doubt probably results from the frequency with which one may follow the birds only to find nothing, or, having drawn a blank, finds a hive with individual effort unaided by the birds, somewhere in a wide area around the site presumably selected. There are so many colonies of bees in Zululand that the possibility of finding one if persistent enough, somewhat clouds the claims for the bird's intentions. Furthermore Zulu folk tales are so often pure exaggerations that one is apt to discount them altogether.

In my own limited experience it often seemed that the birds were searching among the scattered trees of the Zululand game refuges in a very vague, trial-and-error fashion and that they discovered a hive only accidentally. Their exasperating vagueness certainly hints of the possibility that they even may find their assistant hive-opener without any foreknowledge of the hive's whereabouts. If the usual assumption is correct, namely that they already know of a bee tree and are only awaiting the arrival of some creature that can open it, they must have an imperfect memory at best or they possess a most fallible sense of direction. In any case their supposed motive is love of bee grubs and honey and their own inability to get at the food.

That the birds will sooner or later lead one to honey can be verified with a little faith and persistence. The first two or three birds may flutter aimlessly over the countryside and come to rest in a tree around which one may search in vain for a bee tree or an underground hive. It is possible that lack of acute perception or poor eyesight may cause the failure, but to a disappointed victim of an apparent hoax it often looks as though the bird had simply become discouraged by its own inability to locate a hive before becoming fatigued. However, if despite failures each active honey guide is diligently followed, one out of three tries should certainly be productive. Of course this score could be possible even if the birds did not have prior knowledge of the hive—their presumably superior eyesight might enable them to detect the converging flights of bees, or even see them entering and leaving a hive from some distance away.

The fascinating fact of their guiding activities cannot be denied. But a mystery remains as to the evolution of the bird's behavior pattern.

It seems to be generally assumed that the bird's habit originated through a hive-opening association with the ratel, a powerful honey-eating badger-like animal, and that somehow at a later date man more or less accidentally became identified in the bird's nervous reactions with hive-opening. Is it possible that, so far as man's involvement is concerned, the habit might go as far back as *Australopithecus prometheus* or some other hominoid or hominid creature? Presumably the requisite amount of time to evolve the habit has passed since the days of these early creatures, and successive generations of hominids may have merely continued the association. If this guess is correct the association between man and bird would attest to the antiquity of man on the African Continent.

In the Hluhluwe game refuge one honey guide once led me with the usual seemingly aimless wanderings to the close vicinity of a bee tree where the bird suddenly appeared to suffer a partial nervous collapse or exhaustion and drooped silently and almost immobile on a twig some eight feet from the ground. Within a radius of fifteen or twenty yards I could discover no signs of a beehive and was about to write off this experience as more evidence of the bird's vagaries when I sat down to rest. In the deep hush of a warm breezeless day I suddenly became conscious of the buzzing of numerous passing bees and with little difficulty traced them to the hollow tree that formed their focal point of activity.

The hive was located about ten feet from the ground and in a very awkward place if I were to open it and extract the honey. The nearly naked gun boy who accompanied me volunteered to open the hive and get out the honey; he promptly clambered up the tree with his hands and almost equally clutching bare feet. At the site of the hive he was forced to hang with one bare arm around a tree limb beside the hive opening while with the other he hacked at the trunk of the tree with his bushknife. With the first blow of the blade there was a sudden movement in the top

of the tree, and what I had ignored as presumably the wood-carton nest of a common ant, unrolled and started clambering along
a larger limb of the tree toward escape into some nearby bush.
It was a little animal with monkey-like arboreal dexterity—a
much-wanted galago, that small elf, the night-squalling lemur.
Although the only weapon available was a 7-mm Mauser rifle
(about the equivalent of a Winchester 270) I shot the galago
without doing too much damage to the softly furred skin or to the
skull.

During this intermezzo, the attack on the bee tree had necessarily been interrupted. Now the gun boy renewed his offensive.
A large number of the angry little insects swarmed about the
opening to the hive. I cringed as the nearly naked man sallied up
into the growing swarm and I saw bees alight and thrust at his bare
back and shoulders, and engage in their furious attacking gyra-
tions close around his vulnerable and only loosely girt loins. He
soon realized that this was to be a painful job of *uku tapa nyosi*
(rob the bees), and adopted a defense that I had never before
observed. Still hanging by one arm and braced with his feet
against the trunk of the tree he lowered his cupped right hand
and filled it with urine which he then liberally spattered over the
tree trunk near the opening, onto nearby twigs, and into the air.
This maneuver he repeated several times. Whether he gave up
because he ran out of ammunition or because the technique
seemed to be ineffectual I did not learn. Certainly I could detect
no sign of diminution of the ardor of the bees swarming around
the boy.

The classical and more hygienic method of distracting bees
around a hive requires the use of a curiously pungent grass known
by the natives as *umbutambutane* sometimes also as *umbymbym-
shane*. This is the same species of grass that the Zulu women
sometimes use when cooking *amadumbe* (taro), to give it an attractive somewhat smoky flavor. The grass grows from a rather
heavy rootstock or corm-like base, and this part of the plant is
chewed until it produces a copious mixture of saliva and greenish
juice which is then sprayed from the mouth as required. Bees are
stupefied by contact with the mixture according to the natives'

claim, who also believe that although they are not given complete protection they will suffer fewer stings by using the spray.
Without controlled experiments it is impossible to say how efficacious this spray may be. As a small boy I used it many times
until I discovered that patented "smokers" did a much better job.

An investigation of the possible narcotizing properties of this
plant extract might reveal whether it might not only dull the bees
but perhaps desensitize the human nervous system and make bee-
robbing a slightly less agonizing procedure.

After a few well-directed cuts of the bushknife the tree fell to
one side exposing several pounds of pale golden new comb partly
filled with completed and capped cells, the rest still open, as well
as a much larger mass of brood comb and very dark honey. Still
with his free right hand the boy reached into the hollow of the
tree and passed down the dripping masses of broken comb. He
then extracted some of the brood comb, leaving part of it in the
tree and passing the rest down for his own use. The natives firmly
believe that they must be scrupulously fair in this division of the
loot, otherwise they will suffer dire consequences. They leave
brood comb for the larvae and pupae it contains as well as some
honey comb; the honey guides consume wax, honey, and bee
grubs.

The boy invited me to join in the honey feast, and after a quick
mental review of possible urinary-tract pathogens that might
have adhered to the comb and not succumbed to the antiseptic
qualities of concentrated sugar or honey, I accepted reluctantly.
I managed to suppress somewhat my revulsion after considering
the etiquette and the diplomatic consequences of refusing to accept the proffered hospitality. The boy got all the drier brood
comb.

Although I had not been tempted to share in honey collected
after this protective device was employed, I never had such feelings with respect to the honey collected when the mouth spray
was used, even though on later reflection that might have been
the more dangerous bee repellent. In the early days I had learned
to eat bee grubs (pupae) as well as honey, and although the
flavor of the grubs is not unattractive it has no particular appeal

to more civilized palates. We did enjoy another item on the list of the bush gourmet's tidbits, the pupae of a belligerent little wasp known as *qandamatseni* whose paper-carton nests we frequently encountered in most unexpected places along trails in the bush. Even a near approach to these nests brought on attacks by the vicious little wasps whose sting was as painful as that of yellow jackets in the United States. Possibly the game of getting at the grubs of these wasps furnished some of the spice that made them taste so well.

Lest any one contemplates trying either of these items of primitive diet, he should be warned that grubs that have darkened in color, that is, are darker than very pale yellow or practically white, may already have developed a capacity to sting. A sting in the mouth by one of these overlooked maturing wasps or bees can be exceedingly painful and would probably be fatal to any one with even a slight allergic sensitivity.

After carefully leaving numerous fragments of comb in conspicuous places around the site of the bee tree we proceeded to the nearest stream to wash away the all-pervading stickiness of honey—and other residues. The boy seemed utterly unconcerned with either the stickiness of the honey, the tree-trunk dirt that adhered to it, or the primary layer of the mess accumulated as a result of his novel protective spray.

As we headed down the hill toward the stream he explained the reasons for his great care in spreading conspicuous and liberal shares of comb around the robbed tree. He repeated many of the well-known stories about cheated honey guides getting revenge for niggardly rewards by leading men to the black mamba, python, or even a leopard. Although one certainly need not agree on the putative motive that leads a cheated bird to attempt punishment of a miserly person by bringing him into close proximity to a dangerous animal, it would be nonetheless not surprising to discover that a wandering honey guide did at times lead directly to a perilous situation. Most naturalists have been led to harmless snakes simply by following the sounds of birds harrying an enemy of their kind, and one who recognizes and always follows these typical harrying sounds has often been led to

Crossing the Amanzimtoti River in 1900.

The Umzumbe Valley in 1927.

The "pastoral" Umzumbe Valley.

White man's home on hill top; below are native homes in the barren reserve.

Human and monkey footprints.

keys warming themselves on rail tracks.

Arriving at Hluhluwe Rest Camp (*Photograph: South African Tourist Corp.*).

Zebra and wildebeest.

Wildebeest bull.

Zebra partly camouflaged by brush.

At the drinking pool.

Tick birds sitting on impalas.

Above: "Black" rhinoceros (*'ubejane*).
Below: The rare "white" rhinoceros (*'umkombe*).

Wart hog in grassland.

Lion in Kruger Park
(*Photograph: South African Tourist Corp.*).

Red-billed hornbill in Kruger Park
(*Photograph: South African Tourist Corp.*).

Cape ringdove (*ijuba*).

Coqui francolin (*intswempe*).

Coly or mousebird (*ndhalazi*).

Weaver bird (*hlokohloko*),
constructing its nest.

Cemented nest entrance.

Male thrusting food
through aperture.

HORNBILL
(*Umkolwane*)

Male with worm
on guard.

Female with young removed from nest.

Anhinga or snakebird (*ikeu*).

Hammerhead (*tegwane*).

Hammerhead and close-up of nest.

Egg-eating snake (*Dasypeltis*).
View of head before egg is cracked.
Ventral view showing separation of scales.

The toadlike *Breviceps* in inflated defensive condition.

Blueskops showing concealing pattern.

Chameleon in pigmented pattern of fright.

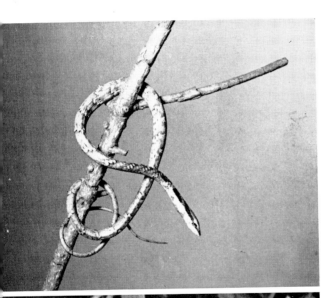

The venomous
bird-eating snake
(*Theltornis*).

Theltornis showing partly
distended throat.

The dread black mamba
(*Dendraspis polylepis*);
shedding.

The harmless file snake
(*Simocephalus*).

The dangerous
puff adder
(*Bitis*).

Chameleon showing degree of eye rotation;
each eye is independent of the other.

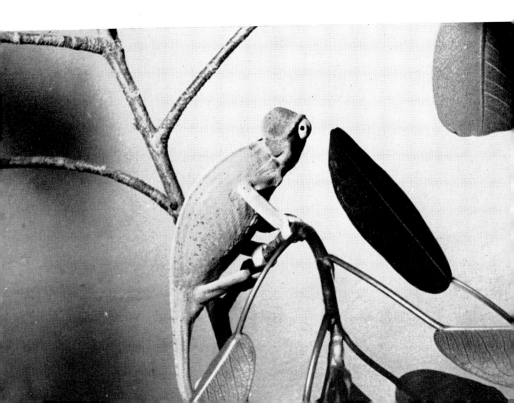

Nuptial flight of termites.

Cross-section through termite hill.

Flask-shaped hatching chamber.

Termite nest.

Cross-section showing
lizard eggs above
fungus-garden cavity.

Close-up of eggs;
white specks below
them are termites.

Monitor lizard hatching.

Newly hatched lizards in artificially opened termite nest.

Captive monitor lizard
retaliating vigorously.

Monitor lizard in
defense position.

Close-up of a Nile
monitor lizard.

Fruit bat showing white ear tufts. Close-up of wing membrane
showing extensive vascularization.

Before and after: Umzumbe Valley in the 1920's and in 1950;
the forest has largely vanished.

A hut thirty years ago; and the tin-roofed substitute today.

Three huts for three wives. Note eroded soil.
Children in a kraal.

Constructing a hut.

The finished hut with thatch roof.

Zulu musician holding
native instrument
(*ugumbu*).

Zulu grandmother showing
native amulets.

Modern native finery worn at a wedding.

Harvesting milo maize for the autumnal beer.

Zulus collecting thatching materials.

Woman carrying pumpkin
on her head.

Woman carrying Dutch oven weighing forty pounds.
Mop effect of woman's coiffure made of red clay and grease.

Zulu in modern clothes, but carrying traditional raw-hide shield.

Zulu in traditional skin-and-fur apron; note modern towel.

A typical Zulu;
note pierced ear lobe.

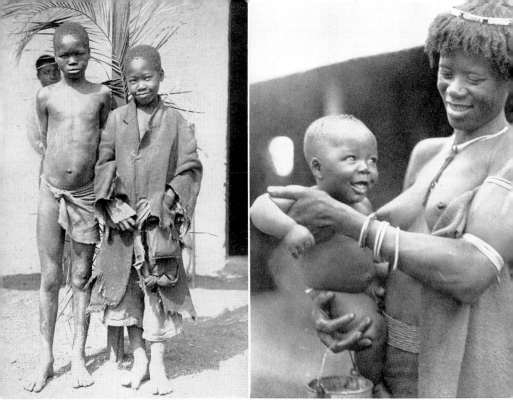

Zulu boys dressed for Easter.   Zulu mother and child.

Christian Zulu women.

cats, owls, and hawks. For this reason it requires little imagination to suppose that with no malice whatever a honey guide might terminate his search for honey when distracted by the sight of an enemy that it or other birds might ordinarily harry. That they might lead a man to close quarters with a leopard would be rather surprising unless leopards are included among the enemies that small birds customarily harry, and presumably this would be too large a carnivore to qualify as an object of harrying.

Weaver and Tinker Barbet

Two songsters, one rare and one ubiquitous, whose voices can be heard throughout the summer are the dark-backed weaver and the tinker barbet. The very attractive notes of the rare weaver have the quality and rhythm of the tinkling of a little music box. They all but play some familiar tune. Not only is the song beautiful, but the little bird that makes it is as attractive as his music. The dark-backed weaver is a little larger than an English sparrow; its back is a dull, bituminous brownish black, and the breast bright yellow. One finds their long, stocking-shaped nests, woven of twigs (not of grass or other fine material, as the nests of most of its relatives) in the cool shadows of the bush. Sometimes their nests are hung from long, pendant twigs of a supple vine that droops over a stream, but just as often they are hung above an opening in the bush. Many weaverbirds hang their nests over water, possibly for added safety from marauders.

(Monkeys, one of the chief dangers to nesting birds, would often be unable to rob a nest in such a position without receiving a thorough ducking. It is possible to test this theory, for the nests are attached to such long, supple branches or vines that one can easily bend them down until they dip into the water. I experimented with a pet monkey, and soon learned that he at least was as averse to wettings as any cat.)

The other songster, the ubiquitous tinker barbet, belongs to those birds whose song seems to me particularly evocative of old memories—perhaps because repetition may be an essential part of establishing such associations. There are few birds whose songs are so persistently repetitious as to become drearily monotonous, and even that most persistent songster of the United States, the mockingbird, has such variety in its song and is so prone to move from place to place while singing that even if he is disturbing when he practices at night, he is seldom accused of being monotonous. The tinker barbet can be heard throughout South Africa in almost every patch of bush or cluster of trees, from dawn to late afternoon. Only where there are no trees is it possible to escape its voice.

The tinker barbet's song exceeds two beats per second for at least four consecutive minutes, and then it may, or may not, take a few seconds' or minutes' pause, and resume. The length of the "break" has nothing to do with the succeeding period of intonation. During these brief interruptions there is usually another bird calling within hearing distance, the two voices thus simulating an off-beat rhythm. The general effect is that of hours of throbbing notes, furnishing a basis for belief that they never stop calling. They seem to be most vocal on hot, windless days when human nerves may already be suffering from the heat, and once a person has taken conscious notice of the song its monotony and persistence can become an obsession. Many people complain bitterly of the throbbing "headachy" songs of the little barbets.

Incidentally, the role of these birds in distributing various species of mistletoe is interesting. Because they are apparently incapable of passing the seeds through their digestive tract they must regurgitate the viscidly coated seeds, which frequently adhere to their beak. Most birds are careful to keep their bills clean, and the barbet is no exception, so it wipes its bill on a convenient twig and the viscid coating attaches the parasitic seed to a new host. Their fondness for the fruit and their habit of planting the seeds, makes a most interesting profitable interrelationship, very similar in fact to the role of our own western silky flycatcher and its distribution of the desert mistletoe from one mesquite tree to another. Without avian aid, many species of mistletoe the world around would long ago have become extinct, or would have evolved some other novel expedient for propagation and dispersal.

FALL AND WINTER

With the end of the summer and the advent of autumn in March, the days gradually become chill and the nights crisp with star-light and moonlight alternating, and with the song of the nightjar piercing the quiet over the lonely valleys of Natal. This is the season of the spirit-soaring African sunrises and sunsets; all na-ture combines to produce a sense of extraordinary, unworldly beauty.

In the daytime there are broadly flaming skies and a diapha-nous bluish haze over the hills and valleys, and the distance-slurred voices of the natives sound lazy and musical, adding to the charm of the countryside. The distance through which sound carries, in fact the whole effect, imparts a feeling of pastoral peace. None other than this trite expression can suffice.

As the warm rains grow cold, and then no longer come at all, with the summer's end the hills turn imperceptibly browner, and the native garden patches of "kaffir corn" become red with a mel-low glow that reflects the comforting warmth of the early morning sun.

Like autumn in any temperate climate, this is always a time of festivity. Until three or four decades ago, there was food in abundance for all, and man was especially well provided for. It was at this season that all day long the well-fed natives once strode over the hills and down through the valleys singing joyfully of what we might call obscenities as they visited from kraal to kraal, preoccupied with the pleasure of sex or talking of the beer drinks to come, and catching up on the latest gossip, which sum-mer with its pressing duties had not permitted them to discuss with fitting leisure. As far as possible, this ancient pattern is still followed, at least so long as the harvests last, but mostly these joyous times have changed, and what may now be said of native life in this one-time season of festival is scarcely worth reporting, except for its increasing drabness. A few ancient customs, folk-lore, and natural-history knowledge persist precariously in north Zululand and increasingly rarely elsewhere; they survived somewhat better on the remote farms where menfolk, women, and children work at tasks set by the white man.

While this was a time of joy for the adult native, for the children there was little surcease from chores, although pleasant ones. While the grain went through its final maturing and drying on the stalk, they watched from early dawn till dark to protect some crops from monkeys, and the red and white kaffir corn from swarms of weaverbirds that came up continually from the marshes and lower valleys where they had been breeding all summer. These birds settled in flocks of thousands in the cultivated fields and in a few minutes might strip large amounts of grain from the ripening seed heads. For some people, the task of chasing away birds and monkeys might possibly have been called work but for the children this was fun. In each field were built small houses to protect the children from any rare inclement weather, and in these miniature huts the guard against marauders was maintained. The huts were set on poles, and such elevated watch towers served to protect the children against roaming carnivores as well as weather. Even long after lions had been exterminated the watch towers still remained, for from their eminence the view of the field below was uninterrupted and the marauding monkeys and birds could easily be seen. Some of these towers are still occasionally used, but year by year their numbers dwindle.

The cozy huts provided a place in which to play, and the work was not too strenuous. When a flock of birds alighted, a few well-aimed stones or throwing sticks was all that was required to scare them away, if only to the next field where they met a similar reception. At each attempt the birds gathered a small share of food, but if the boy watching the fields was skillful with his sticks and stones, he also took a toll of food for himself, the birds he killed adding to his menu. The owner of a rubber slingshot was a prince in those early days. The birds, roasted over coals—feathers and all—and eaten whole like sardines, were delicate morsels to his unjaded taste buds. Even the head was eaten, and the nicely charred skull crunched with what seemed, from watching the boys, to be highly satisfactory results. This was gourmandism at a primitive level indeed, but it was clear that the finest food savored by the most exquisitely trained palate, could provide no greater pleasure.

The birds that most frequently fell victims in the fall, in addition to the weaverbirds, were fat *uzwaubes*, red-shouldered widow birds, which look almost exactly like our American red-winged blackbirds, and the long-tailed *jobelas* with their short-tailed, dull-plumaged mates usually called *i'ntakas* by the Zulus. These were all well-fed birds, present in hundreds and easy to get. Although they are not large, most of them being about the size of a house sparrow, they seemed to be eminently satisfying food for the small guardians of the harvest.

Nectar Feeding

During the dry season winter-flowering plants for some birds supply both food and water, an interesting demonstration of avian adaptability despite the normal inflexibility of their behavior. The use of nectar for its water content may at times become even more important than its nutritive value because by this expedient relatively sedentary species can occupy areas that are rather far from permanent water and yet continue to forage for food in their own locale. Seemingly this would reduce concentration near streams or springs and so somewhat reduce competition for food. Whatever the cause of this extensive adoption of nectar feeding by species that are not specialists in the art, the habit is interesting.

There are many specialized nectar feeders, birds that are ecological counterparts of our own hummingbirds, but they belong to a different order than that of the hummers. The male sunbirds (*Nectariniidae*) bear the same general plan of iridescent colors and mixtures of greens, purples, reds, and other vivid colors that make male hummers so gaudily notable under some conditions or inconspicuous amid their floral chalices, but the sunbirds very rarely hover in mid-air while getting their food, and they prefer to perch on or beside the blooms.

During my latest trip to South Africa I observed a notable display of flowers and birds when a solitary red-flowered *umxabo* or *uhluze* tree (*Schottia brachypetala*) became alive with multitudes of brilliantly colored birds feeding from the blossoms that appeared in winter. There were scores of the scarlet-chested, black-bodied sunbirds in the tree, and at any one time their continual comings and goings and nervous hoppings from limb to limb kept the tree vibrant with life. The birds were so numerous that even some of the adjacent thorn trees seemed to emit a continual twitter that was always most melodious, and the entire scene was lively in the extreme as contrasted with the unusual winter drabness elsewhere. Many other species of birds were also present, including the black-capped bulbul, spot-backed weavers, and black-headed orioles. At each visit it was pleasant to sit in the warm winter sunshine enjoying the balmy, vagrant breezes, the aromas of the bush, and quietly watching these birds. Once I found the glowingly colored body of a black and scarlet sunbird that had fallen dead, but which still shone in all its external plumage brilliance. High in the sky, each soaring above the other until the farthest was a mere pinprick, were the vultures, symbols of death, but around me was the life of the bush, which is to say life for some and death for others so that life might continue.

Not least of the symbols of this universal phenomenon was the dead-looking tree before me with blood-colored blooms overflowing with nectar derived from the unseen sap. The very silence of the place, other than the twittering of the nectar-feeding birds, was emphasized by the audibility of a leaf rattling in the treetop, or the movement of some animal in the surrounding bush, or the raucous calls of the white-bellied crows which, like the vultures, are living sarcophagi.

Use of nectar by birds that are not detectably specialized for this habit is notably common in Natal, more so than I would guess it to be in the United States, hence extension of the feeding habit to include introduced plants is not surprising.

Eucalyptus trees, or blue gum, were once a rather useless type of vegetation for birds in Natal. For most birds these trees, so widely planted throughout white-occupied areas, provided little attraction except for roosting sites and, for very few species, nesting places and some scanty shelter. This was about the extent of hospitality toward animal life because the bare ground surfaces under groves or forests of the Eucalyptus and wattles offered little inducement for any birds to become residents in plantations of these foreign trees.

To an extent this still holds true, yet the nectar from the heavy blooming and extensively planted blue gums now supplies food and drink to many species of birds. This may have been particularly true in the very dry year of 1953, the seventh year of drought, during which water was at a premium for birds as well as for other animals. Many weeks passed without any rainfall, and the days were warm and usually windy, so that for the truly arboreal birds, those that are accustomed to getting water from raindrops or dewdrops on the foliage, the only available source of liquids seemed to be found in nectar.

My recollections from some thirty years earlier suggested that many species of birds which I had previously never observed using nectar from the Eucalyptus trees had apparently learned to do so some time within the past few decades. Among the previously unnoted species of birds that have adopted this nonnative source of combined liquid and nourishment were the spectacled weavers, black flycatchers, mannikens, the striped colies, and the white-eyes. Spot-backed weavers and black-headed orioles were apparently early users of Eucalyptus nectar. Some of these birds undoubtedly have been accustomed to seek nectar from numerous native plants including the aloes, which also blossom in winter.

During the dry months the spot-backed weavers (hlokohloko) assemble in huge flocks, and then their use of the Eucalyptus becomes conspicuous. Day after day, hundreds or sometimes a thousand flock into the tops of Eucalyptus trees in bloom. These birds seem to have been among the first to exploit the Eucalyptus nectar. My first observations of this habit date back more than thirty years, but there seems to have been an increase in dependence on this introduced source of supply, possibly as a result of recurrent droughts. These birds also feed on grains, and they are forever dropping down into the fowl runs and fighting among the chickens for the cracked corn scattered in the yard.

When at length the Eucalyptus trees begin to bloom, the flights of spot-backed weavers, a minor avian spectacle, occur usually just as the sun rises disc-high at dawn. Day after day with a variation of only a few minutes, several hundred, sometimes a thousand or more of these gregarious bright yellow or olive (male and female) birds begin pouring into the nectar-laden trees, and for the following half hour the tops of the huge blue gums are aquiver with bird activity.

The weavers' bodies at this time of year are perfumed with the delicate odor of Eucalyptus blooms, and in spite of the drought and the hardship that it is working on many species, at least these birds remain fat and seem to be in excellent health.

The dietary adaptability of the spot-backed weaver is also displayed by its catholic taste in food, which includes new buds of awakening plants and seedling corn, peas, beans, and other crops. Although it cannot be proved specifically, their behavior seems to indicate that they also scrape scale insects from the twigs of trees. At times I watched a half dozen of these birds as they scraped material from the twigs of a golden flame ("shower of fire") plant, and since these young twigs are covered with lenticel-like scars that very closely resemble scale insects, and since there were also some insects present, it appeared that these birds were actively engaged either in pursuit of this source of food or were deceived by the markings of the bark.

The Eucalyptus trees are among the first sources of food visited in the early winter morning. At this time the blooms are so loaded with nectar that a vigorous shake will scatter droplets of sugary liquid in all directions.

Soon after the birds have worked through the Eucalyptus blossoms, the flocks spread over nearby gardens where in winter they devote special attention to the blossoming Euphorbias on whose long spikes the blossoms are so arranged that the opening of each bloom is pointed toward the ground. While feeding on the nectar of these flowers, the spot-backed weavers must stand virtually on their heads to extract the liquid. Their fairly short conical bills are certainly not adapted for nectar feeding, but they probe all blossoms that may contain nectar. Apparently they do only minor damage to the blossoms and are probably as effective in pollination of some species of flowering plants as insects or the sunbirds with their specialized nectar-gathering habits and structures.

By June true winter has arrived and the nights are almost always crystal clear and not infrequently frosty. This is the season of winter-flowering plants, of warm but tempered sunshine, dry breezes, and miles of pale tan grasses that look like past-ripened wheat. Many trees shed their leaves, even those that like the crimson coral tree will soon put out their numerous clusters of colorful rosettes. On the breezeless nights when the sky is clearest, cold air from higher elevations flows downward in silent and invisible rivers to form pools or lakes in the valley bottoms, where sometimes the pale light of early morning reveals milk-white "frost bowls" in these cold traps.

When the moon is full the countryside is glorious in the silvery heatless light, and the jumble of hills are etched into sharpness by the black shadows of bush and valleys. It is at this season that far and wide the nightjars toss their clear notes to each other in a chorus that gradually fades to bare perceptibility in the distances. Their songs have a quality and rhythm of notes quite similar to that made by its relative, the whippoorwill. The notes are cheerful, but in the infinity of space and the sublime beauty of a perfect night, they suggest sorrow.

One June night I chose a moon-tinged eminence near the mission from which I could see and hear the world around me. From a black gash of bush in the valley set in shining hills, an owl called again and again, while from a greater distance a dog barked and a rooster crowed, their domestic familiarity suggesting the unbelievable presence of other communities and other human beings.

The shadows lengthened as dawn approached, and the hills threw great arrowheads of blackness across the valleys until they reached the dark gashes of bush that filled the ravines between the hills. As the moonlight failed, a faint line of gray outlined the horizon. With the nearing of dawn, the night sounds increased momentarily in volume, only to fade again before the threat of daylight. As the cool, gray light of dawn diluted the night, the day birds wakened from sleep, and on all sides was the rustling of birds exercising their wings after a night of immobility. Here and there a bird chirped as increasing daylight brought greater confidence.

Soon after the first birds started calling, the tambourine doves (isibelu) sang their song. Over and over again came their sad-sounding refrain, which the Zulus have translated to mean "my father's dead, my mother's dead, my sister's dead, so my heart goes do do do do do." Soon a pair of them waddled up a path, stopping now and then to pick up food or gaze about them Each looked black and white in the dim light, black on the back and white down the front, comical despite their sad song, carica-tures of stout men-about-town in evening clothes walking home after a night out.

With the appearance of the doves, day seemed to have come. Then, above the disjointed medley of twitterings, rose the clear, flute-like call of one of the mystery voices of the bush. It came from near at hand, and yet the songster remained invisible. To see more than a foot or two into the dense tangle of brush was impossible, but a trick I learned in America worked magic even in Africa. By placing the hand against the mouth and imitating the squeak of a distressed fledgling, I got an instantaneous re-sponse. From the maze of vines and branches, a gorgeous bush shrike came into view. The back was a brilliant green, the throat and upper breast bright scarlet, and below the scarlet was a bib of jet black, bordered below by a thin margin of red merging into the yellow of the breast to give a shading of orange. This was the four-colored ivovoni of the Zulus, a lovely final reward of my dawn watch to remain in mind as I walked back to the mission.

There are many beautiful species, but few other members of the shrike family are more generously endowed with color than the four-colored bush shrike. Other species, such as the spook bird or spookvogel or southern gray-headed bush shrike, are at-tractive chiefly because of their rich green and gray plumage. The spook birds sing with a long, doleful note somewhat like the sound produced by drawing a file across a big saw. This note is often followed by other sounds, usually a dry, unmusical clacking noise, so that altogether the common name is not inappropriate.

Although the spook birds, like most shrikes, are largely insectivorous, they also catch larger prey, and a locally applied name sometimes used by small boys, "chameleon catcher," is at least partly descriptive of their food habits.

On three occasions, when I was trying to trap genet cats, I caught spook birds in traps baited with rats. Each time the rats were in various stages of putrefaction, yet were attractive enough to bring these tree-living birds to the ground for food. It may have been a coincidence, but the species of rat used as bait was the same, a bush-dwelling variety which when fresh does not have the usual objectionable ratty odor. I used other species of rodents more often as bait because of their greater abundance, but in no instance were these effective.

Many South African shrikes are ordinarily observed in pairs; and in some species, one of the most interesting characteristics is that of call and response in the songs. The most persistent call-and-answer habit in the bush is that of the bou-bou shrike, the *uboboni*, a rather dull-plumaged bird, living in the densest part of the bush, not far from the ground. I found them almost invariably in pairs, and they called and answered in such immediate succession that I often thought them to be one individual with two different song phrases.

Flocking

The earlier-mentioned watching of large flocks of weavers in Eucalyptus trees gave rise to my interest in the phenomenon of flocking with its related questions. Africa is certainly the place where a study of behavioral activities might explain the origin and evolution of gregariousness and, ultimately, of social behavior of cohesive flocks as differentiated from looser aggregations. I cannot claim to have found answers to the innumerable questions that arise, but in the course of my observations I made a few speculations that may be of interest.

For one thing, I had opportunity to note the banding together, in loose unorganized flocks, of many different species of birds, representing several families and as many diverse ecological niches. In winter, when the reproductive drive was over and the arduous duties of nest building, rearing the young, and molting had ended, I noticed these flocks of different species moving together, drifting back and forth through the bush, each gleaning its special kind of food in conspicuous amiability. The breeding-season fixity of ecological niches seemed relaxed, and only general-habitat requirements seemed to remain compulsive. Sometimes this companionate grouping appeared to result from an accidental association, a premigration adaptability, yet there were many groups composed of birds that apparently did not migrate.

On blustery days, in the sunny spots sheltered from the wind, such aggregations were not uncommon, but even on quiet days these flocks moved through the bush, loosely keeping together, occasionally breaking up, and reassociating again with other individuals elsewhere. There seemed to be some sense of mutual comfort and protection afforded by the greater numbers, and when a somber bulbul flitted across an open space to another patch of bush, quite often one or two members of a flock of white-eyes (greenish, kinglet-sized sprites) took flight also. Whenever these small birds flew, they chirped shrill call notes, and one by one the entire flock followed the lead of the first. The Zulus call these birds *mehlwana* (little eye) in reference to the white eye ring, or, more characteristically, *nobutukwana* (little scolder), because they are invariably present at all snake- or hawk-harryings, scolding vociferously. When these birds take flight, other, even unrelated, species in the aggregated flock are apt to follow. There is apparently a mildly contagious stimulus arising from the sight and sound of movement.

These flocks probably change in composition during a period of time, but it is noticeable that whenever one encounters such a group, at other nearby points the bush may be quite empty of bird life. The groups will hold together until breeding and "homesteading" (in spring).

While watching these aggregations, my special attention was focused on species-flocking habits, and those could be studied best among the mentioned weaverbird family, including the whidah finches and waxbills. The most notable example of crowding, the buffalo weavers, has been described in several books. Their nests are so closely packed together that they construct virtual apartment dwellings. So intimate a group could almost qualify as a social rather than a merely gregarious assembly. From this extreme the weavers demonstrate descending levels of mutual tolerance or dependence, through the loose tree-colonizing by the spot-backed weaver which makes numerous distinct but nonetheless closely spaced nests in one tree, to the colonies of thickbills and other reed-building weavers, or small flocks of the parasitic pin-tailed whidah *hlekwa* (usually a male and five or six females), to winter flocking but breeding-season isolation, to the presumably year-long pairing and solitary nesting of the black-backed and other weavers.

In addition to these examples drawn from one family of birds, there are others of interest such as the colies or mouse birds, starlings, pigeons, hornbills (the latter possibly family groups), and the loose flocks of babblers and woodhoopoes.

The habit of flocking is so widespread that it would seem to provide survival advantages in spite of the problem of finding food and shelter in sufficient quantities for so many birds. It may be that flocking is of value chiefly as a means of defense against predators, because in effect a flock of birds represents an organism with a large number of eyes watching for danger in all directions at all times. In the enormous winter flocks of the spot-backed weaver it was obvious that in one respect at least (availability of nesting places), the flocks would be unable to remain together in the numbers I saw in wintertime.

Apparently, these winter flocks are not merely an accumulation of individual birds, but possibly represent an assemblage of independent flocks. At times behavior illustrated this structure, for toward the end of the winter season there was an increasing tendency for segments of the main group to break away and maneuver independently. It would be interesting to know whether the flock components of the huge assemblage retain their identity throughout the flocking season, or whether these smaller groups that break away are a random assortment of individuals that happened to be nearest when the break-away took place.

I could appreciate the advantage of flocking when I once saw a hawk appearing in the vicinity of the birds. There were many hundreds of birds feeding on the top of a Eucalyptus and making a terrific din, then apparently one or more individuals gave an alarm signal. Instantly the call was repeated by others, then the majority fell silent and watchful. When the hawk plunged toward the tree, the silence was broken by the murmur of hundreds of wings as the birds made a concerted escape, dropping first toward the ground and, having gained speed, darting away from the danger. The dangerous little hawks do not seem to pursue the flocks once they have taken wing.

Despite their friendly gregariousness, desperate encounters take place between the weaver males. Although the weavers are primarily arboreal, they frequently feed on the ground, and possibly for this reason seem to have no fear of continuing on the ground a battle that started in the trees. Their fights are not mere bluffing affrays. The males come to close grips, and they lie on the ground pecking and clawing fiercely, meanwhile rolling over and over, and then lie still, feeble from exhaustion. These combats become noticeable as early as the break-up of large flocks but before serious nest building and oviposition occur, a fact that suggests the value of studying specific endocrine activities and the general gonadal cycle of these birds. Some time before actual nesting, male weavers will often engage in serious fighting.

It is amusing to observe the groups of spectators at these fights, usually a dozen females, who would stop feeding in the trees in order to gather close about the ground to watch the combat. Wherever the struggling males roll, the females hop about in a half circle, intently engaged in awaiting the result. After minutes of struggle on the ground, one of the combatants usually manages to break loose and take to flight. However, the remaining male, if he can do so, inevitably engages in pursuit. Often both birds become so exhausted that in their flight they resemble fluttering butterflies rather than the very vigorous flying birds they normally are. If the fugitive male manages to gain on its pursuer, the flight may continue even up to the concealment of thick foliage where both birds momentarily disappear. However, a fight is by no means over when one of the battlers has been defeated and has attempted to escape. Apparently with these birds a fight is a desperate encounter, a battle to the death if possible. In a gregarious species, it seems extraordinary that fights of such vicious intensity should take place before the breeding season.

So well developed a trait logically should have survival value to the species even though it may be lethal to many participants. Does the value, if any, consist only of selection for vigor and skill of reflexes? The high pressure exerted on males throughout this combat period must tend to select the most vigorous and pugnacious breeding individuals, but since the species is polygamous and courtship is largely limited to demonstration at the nest openings, vanquished birds that have survived serious injury must later steal their opportunity to mate, even at the risk of encountering the dangerous punishment that the original victor is capable of inflicting.

Clustering

Not many years ago biologists believed that no birds could habitually or regularly become comatose from cold nor in any sense hibernate for periods of even short duration. This conviction had resulted as a reaction to ancient beliefs about the annual migratory disappearance of birds during winter.

Recently, however, especially following the discovery of hibernation in the California poorwill and later studies that described the surprising fact of nightly drops in body temperatures of hummingbirds, it seemed quite probable that semi-coldbloodedness, or at least well-defined temperature changes might be expected in other species of birds. It is for this reason that I gave special attention to a common species of the colies or mouse birds (ndhlazi), a small family of long-tailed, sparrow-sized birds (*Coliidae*) reported to have drops in body temperature. Their clambering through the trees is often more mouselike or even reptilian than avian, and with their long tails and their slithering about in the branches they are also remotely suggestive of parrots.

Weeks of observation in winter with air temperatures falling as low as 44° F. revealed little information that seemed unusual except for their almost lizard-like love of the sun, the long periods spent in basking, and their tendency to cluster together with their bodies touching.

These colies were the so-called striped or speckled species with wide distribution extending from the Cape northward to the Zambesi and Nyasaland. This engaging little family of long-tailed birds is confined entirely to Africa. The speckled coly is common—in fact, it is too common in the opinion of residents who attempt to keep a few fruit trees or to raise winter vegetables.

The common coly is one of the most silent birds and seems to have no song in the regularly accepted sense. Its loudest notes are wholly unmusical and limited to feeble buzzing calls, scarcely louder than a whisper. The coly also has a rather faint alarm and signal call that is given when a flock is in danger or is about to fly to another tree. These signals are heard best from a photographic blind when the birds are often within a few feet of the listener. The colies seem to congregate in small flocks of six or seven, and they communicate with each other by means of those buzzing notes. Before leaving a roosting place, the rapidity of the calls and their loudness rises from bare audibility a few feet away to a louder, brief crescendo as they take wing. Almost invariably all members of a flock either leave together or in tandem fashion, and if one of the flock happens to be left behind it shows distinct symptoms of nervousness. Before taking flight they invariably "talk things over" in their buzzy subdued stage whispers, then depart in ones, twos, or larger numbers. In flight, their long tails and stubby wings give them a striking resemblance to broad-headed arrows, and when a flock is silhouetted against the sky they suggest a massed "flight" of crossbow missiles.

Although they almost always remain in small flocks, just before dark it is not uncommon to find two or more flocks joining together before flying to some secluded spot to roost. I was never able to follow these flocks to their sleeping place nor to learn whether both flocks combined in one group or whether they separated so as to maintain their identity. I saw combined flocking only during the winter and I surmise that this was done so that twelve or fourteen might huddle together for additional heat conservation. This would be a logical expectation because they would benefit by forming a larger mass with a smaller proportionate surface area, a factor in conserving body heat for a longer time. Individual flocks, six or seven birds, definitely huddle together in a dense mass while sleeping or during rains, and this habit and their devotion to basking were the chief visible concession to a reportedly poor capacity to generate body heat.

When they bask, they often remain in the sun for half an hour or more doing nothing beyond soaking up the gratifying warmth from the sun's rays. Their slightly yellowish-brown breast and belly are covered with a scanty, fur-like feathering which they almost invariably present toward the sun while they bask. As they cling to branches close to each other there is a good deal of mutual preening, such as seen in lovebirds, and they are addicted to cuddling close to each other.

In contrast to the colies, practically all other diurnal birds exhibit almost constant activity, which means expenditure of energy and heat production. They are continuously hopping about, flicking wings or tail, turning their heads, preening, or going through other motions. In comparison the colies appear sluggish, and they often remain quiet for half an hour at a stretch; preen only sporadically and do not even perch as do all our familiar birds, but cling with feet elevated as high as the level of the wrist joint (the bend of the wing) or even as high as their heads. In fact they actually hang to their "perch" rather than sit on it. Perching birds benefit by an automatic device that closes their feet tightly around their roost as soon as they relax and squat for the night. The colies must have some quite different anatomical construction to allow them to sleep without continually contracting their feet and leg muscles.

While basking, the birds appear to sleep, at least they hang motionless with eyes shut and with every appearance of sleeping, but there is always at least one merely drowsy-looking individual whose eyes periodically open wide enough to scan the surroundings. When it falls asleep or at least closes its eyes completely, another of the flock takes up these casual sentry duties. It is always a scant few inch dive or slide, from basking places to shelter in the recesses of the tree or bush in which they have been resting.

When one of the earlier-mentioned voracious little hawks is in the vicinity, the colies as well as all other species of birds, are almost continuously alert, nervous and flighty, ready to dive for cover at the slightest sign of danger or in response to the alarm call of any other species of bird.

Because the colies' habit of clustering together under adverse temperature conditions might be one of the symptoms of their supposedly poor capacity to generate body heat, I spent much time in looking for their roosting places at night, or for an opportunity to find them caught by a sudden rain. Ordinarily the birds took flight at the first pattering drops of a shower and disappeared to some undiscoverable favorite place of their own. But on the occasion of a rare winter thunder storm at Umzumbe mission the drops came too suddenly, and I could see the birds indulging in an extraordinary proceeding.

At the first few drops of rain I moved out to a porch near to a fruiting tree that had attracted the local flock of colies. Despite the initial scattered drops, the birds continued basking throughout the brief remaining moments of the sun. Some rare optical phenomenon, one I had never previously witnessed, caused a halo to form around their bodies in the prethunderstorm light amid the few droplets of rain. As the rain increased in intensity one of the birds flew to a Bougainvillea bush, undecided whether to fly away to another refuge or remain. But when the rain turned into a downpour, a second bird joined the first, and both hung with feet almost even with their bills, "billed" each other, and snuggled closer with heads touching and tails crossed. The two were almost immediately joined by a third, which then hung breast to breast with the first couple, and a fourth flew in and settled on top of the first three, squatting on them in such a way that its feathers shielded their heads. As the rain continued, this bird sank lower and lower, others joined these first ones, until all but one were clustered, presenting a virtual shingled shield of birds, with their backs, that is, the water-shedding nonfurry side, presented to the rain. The remaining bird gave a preflight call and headed for the dense foliage of a nearby tree, where the others soon followed. Doubtless the clump was then re-formed within the concealing shelter of the tree's foliage.

On only one other occasion did I observe such a cluster in the rain. It was unfortunate that both times in the dim light under heavy storm clouds it was impossible to obtain photographs.

For several months I looked in vain for clustered sleeping flocks. Just after sunset each day these colies used to fly up to about the middle of a fifty- to seventy-five-foot cypress near the mission, and from that point hop and crawl forward to near its top. Frequently a second flock from an adjacent area joined this group, and five to ten minutes after sundown in very cold air (about 45 to 50° F.) they would all take flight to a roosting place far around the bulge of a nearby hillside.

Once, as I attempted to get a better line of sight on these flights, I watched the birds as they gathered and waited for the second group, the one moving the greater distance invariably being ten minutes late. As they climbed upward toward their usual take-off place, either the presence of an observer or the coldness of the night changed their usual behavior, and all came to rest near the very tip of a tall cypress. There they remained until dark had settled. At 9:30 P.M., thinking that the birds must by then be comatose from the cold, I cautiously ascended the tree. At the start of the climb I heard a faint coly alarm call, but from that time on the birds made no additional sounds nor was there any fluttering indicating that they had left. I climbed as far as I could, but when I saw no birds came down again.

After weeks of watching I tried again. In the pitch dark of a cold cloudy night, using the flashlight as little as possible, I cautiously climbed the tree to where once it had been split by lightning and then healed and grown still higher. Since I had carefully watched until total darkness I knew that these strictly diurnal birds had not left their hiding place. Also an occasional faintly whispered call note proved their continued presence. Beyond that, all birds remained utterly silent while I reached the highest attainable point. Their hiding place was three or four feet above my highest reach. When I could not see any birds by flashlight, I gently shook the tree, then swayed it more and more vigorously. But the birds did not respond with the usual give-away alarm note. Finally the only remaining method of finding them was to reach upward as far as possible in order to feel through such clusters of needles as I could reach, but the tree was too slender at the tip, and it started to bend and sway outward. There was nothing to do but make a quick, desperate slide down into the blackness below, where fortunately the increasing rigidity of the tree stopped my outward descent. Even with all this commotion there was no sign of the birds.

There is abundant testimony that nighttime huddling of colies is a customary manner of sleeping during the winter months, yet I could never personally observe nor photograph them while clustered. Since I was not able to capture any comatose birds and take their body temperatures, I decided to trap them and in this way obtain temperatures under various conditions. I wished to get direct information of their reported cold-bloodedness.

Because of the scarcity of flocks in the neighborhood and because I did not want to disturb those that were available for field studies, I offered a native a reward for capturing, unharmed, two members of local flocks. At the same time I requested him and other men to broadcast the information that an average month's wage would be paid for information on the whereabouts of an accessible roosting place, a cluster of rain-wet birds in a photographable position, or a nest of the birds. No bird was to be killed.

In spite of the reward, the natives for a long time did nothing. My request for just two unharmed birds seemed to be beyond the efforts of the local natives—perhaps because I insisted on their use of the relatively humane standard cow-hair slip noose. Eventually, however, one herder boy, using steel traps, caught two colies and brought them to me. Each of them was injured so badly that it was necessary to put them out of their misery as soon as possible, but before doing so I took their temperatures under several conditions.

The first bird had suffered a broken leg, but within a few minutes of its capture, soon after its feeding had started and its belly was full of fruit, its body temperature was 38° C. (100.4° F.) while the environment was only 12.5° C. (55° F.). Although this seemed to indicate little tendency toward cold-bloodedness, 100.4° F. is low for birds in general especially after vigorous exercise. To test the effects of inactivity, I enclosed the bird in a plastic bag to discourage struggling and kept it in the dark at an environmental temperature of 22° C. (71.6° F.) for two hours and a half. Unfortunately defecation and a good deal of moisture dampened some of the body feathers, but the end result in that short time was a body temperature of 24° C. (75.2° F.), just the effect that would be expected if the species was a heterothermic (intermediate, between cold and warm-blooded) type.

To test for possible effects of shock in this lowering of temperature, I placed the bird in a small cage in full sunlight for fifteen minutes. It wakened immediately and hopped and fluttered about, seemingly normal except for the injured leg. In another fifteen minutes the body temperature had risen to 39.2° C. (103° F.). Continued fluttering started the bird panting after only ten minutes, and its body temperature was then 42° C. (107.6° F.). Its feathers were then dry. Again I returned it to a dark place at 22° C. (71.6° F.), and after an hour it had a temperature of 40° C. (104° F.). By evening the bird's temperature had fallen to 36° C. (96.8° F.) and the following morning, after being awake for two hours, it was 35° C. (95° F.).

These are by no means satisfactory data but under the circumstances it seems probable that the birds are heterothermic, that they can and do experience cold torpor probably as a fairly regular and normal physiologic phenomenon, and that their activity temperature is rapidly attained but that it tends to remain around 37° to 38° C. or about that of human beings, low for a bird. In some respects they may resemble some bat temperature changes, cooling rapidly and then as rapidly generating heat when needed.

Of all the birds I watched, the colies seem to depart most widely from the expected and fairly standard characteristics of more familiar birds. Their flocking is distinctive in many respects, including of course the clumping while sleeping. Their mouse-like clambering through trees has already been mentioned. Also, they have the habit, unusual among South African birds, of bathing in water or dust, whichever is available.

The variety of foods in their diet is particularly unusual. I saw them snipping out segments of the leaves of the passion fruit (*Passiflora edulis*), and they will practically defoliate a vine of this plant. After feeding on these leaves they may fly to a blossoming aloe and take triangular bites from the lower edges of the somewhat fleshy petals, cutting back clear to the ovary, and they appear to consume the petals and the nectar as well. From here they may repair to the blossoms of the Eucalyptus, where they also feed and apparently drink the nectar, then to the ground in a patch of sprouting peas or beans, so that a small plot of vegetables will be completely destroyed. After feeding on vegetable matter, parts of their digestive tracts are distended with masses of green material. The most extraordinary dietary trait is their apparent total avoidance of insect food. Although I watched them carefully for many days at a time, I never saw them eat insects of any sort, and, most surprising of all, they even appeared to ignore flying termites, a source of food that is almost universally relished by birds, mammals, reptiles, amphibians, and insects.

Unlike many gregarious birds, the colies do not seem to be quarrelsome. When they feed on some small fruit like a loquat, two birds often lie side by side gradually consuming an entire fruit with no sign of greed nor haste to get the lion's share. And if food is scarce, a third or fourth may join in, one sometimes alighting on the back of the first comer, where it may continue to perch while sharing the fruit with others. Even under these conditions there is never the slightest indication of resentment.

Colies do not extend this amiability to other species of birds and there are frequent occasions when the ubiquitous "toppie," *ipotwe* (the black-capped bulbul), attempts to snatch a piece of fruit from its somewhat smaller competitor of the bush where both must subsist on a limited supply of food. On these occasions I have seen the colies rearing back until they were balanced on their wings and long tail, legs free, and claw simultaneously with both feet at the offending interloper. Even under the impetus of anger and self-defense I have not seen them use their beaks against the toppies although the slightly parrot-like bill is sufficiently powerful and sharp to make clean cuts through fairly tough leaves, and to administer painful pinches to a human hand holding the bird.

Most endearing and amusing of their habits is that of the male and female lying side by side while jointly incubating their eggs. This habit had long been reported by natives and a few white persons, but by some mischance in many years of work I had never found a nest of this species until the last day before I left Natal when I discovered an incubating pair on the rather shallow saucer-shaped nest, bodies tucked closely together, with their long tails protruding beyond the confines of the twig-constructed platform.

TRAPS

In my years of studying birds I have increasingly relied on watching and photographing them rather than collecting specimens, although I have also hunted them ever since the slingshot days of my early childhood. On many occasions, when it was necessary to obtain specimens, I have enlisted the uncanny skill of the natives, especially the young boys, pressuring them—with spotty success—to use more humane traps, if any trap can be called that.

Bird catching has always been a universal sport among the young Zulus. While today commercially produced steel traps are in use, in the early part of this century many kinds of self-made traps were common.

The catching devices that children then were accustomed to use were mainly the *umgogo* (twitchup), the snare hung in a runway, lime, throwing clubs, and—most deadly of all native methods—the *isifu*. This bone-crushing device was somewhat like a figure-4 trap, or dead fall, but with a flat, heavy stone instead of a box or log.

For the past thirty years at least, most boys have managed to obtain an occasional sixpence, and with such wealth buy regulation steel rat traps. These have been wickedly effective, and each year they have accounted for large numbers of francolin and other ground-feeding birds. The youngsters were and are extraordinarily clever and industrious at this trapping. For most of them it was often their only source of meat. They weeded long trails at the edge of old fields, and along these trails they scattered grains of corn, "mealies," or kaffir corn. Where the steel trap was set, a fence of twigs four to six inches high was built at right angles across the trail to make sure that the birds did not walk around the trap, which was partly buried in the ground in the middle of a small gateway.

This use of the steel trap in the trail-and-fence combination was a development of the ancient methods employed in setting up the *umgogo* or twitchup. Two variations were used in making the twitchup trap, depending on the results desired. If the bird was wanted alive, the noose was made in the form of an ordinary loop like that used in tying a square knot, with the second turn of the string omitted. The short end of string leading from the noose was attached to a stake that was driven into the ground, and the long end of the string, usually about eight or ten times the length of the short one, was attached to the sapling that acted as the twitchup. After the *umgogo* was set, the loop was carefully concealed under a sprinkling of dust. When a bird stepped on the release mechanism, instead of having a slip noose snatching it around the neck or body, causing

sudden death, this arrangement of the noose tied the bird by
one or both legs holding it firmly without injury.

The plain snares used by many boys were invariably made
from three hairs extracted from a cow's tail, the hairs being
twisted together into one long strand. The hairs from a horse's
tail were not so desirable. They are brittle, seem to be no stronger,
and are harder to work with. To be most effective, superstition
demanded that the hairs for the nooses be filched from a living
cow, an exciting process, for there was an essential ritual connected with their extraction, and the cows objected strenuously.
They dashed away at full speed, kicking and cavorting, while
the ragged little herder boy hung onto a handful of the hairs
and tried to expectorate accurately enough to place some saliva
at the junction of hairs and tail. The saliva was reputed to en-
hance the virtue of the noose and to make the operation less painful for the cow. The former reason was probably the only motive,
the humanitarian reason a camouflage. The requirement certainly
made the process of extraction more exciting for the boy.

The hairs were knotted together at one end, three to a bunch,
and then each bunch was twisted separately. The knotted end
was firmly held while the other ends were placed about an inch
apart and pressed against the boy's naked thigh. Before twisting,
the leg and the cow hairs were again lubricated with saliva, then
the hand, the palm down and pressing against the hairs, was
moved forward as far as possible so as to twist each hair without
allowing any strand to escape rotation. While this was being
done, the knot was still held firmly, and the final result was that
each hair was stressed and twisted but all were still separate.
Following this step, and still holding the free ends, the knotted
end was allowed to rotate, and this rotation turned the hairs
together, making a single strand. The entire operation took only
a few minutes, and the strands would not unravel when wound
in this manner. An ordinary slip noose was tied in these elfin cow-
hair cables and the completed product was used in various ways.

A favorite method of employing the snares was to build a trail
similar to that used in setting the *umgogos*, except that with

snares the fence was bridged across the trail in the form of an arch, and one (or more) of these snares was tied to the top of the arch and allowed to dangle into the trail. While feeding, a bird normally progressed along the guiding trail until it thrust its head through the noose, which to it must be indistinguishable from impeding grass. When prevented from moving forward, a bird would back up to escape the obstruction or try to find another path around the restraining strands. Once the bird backed up it was captured, for the fine noose would slip under the feathers, and since it was then pressed by the surrounding quills, it could not spring open as it would naturally do when released from tension. A few violent struggles and a bird caught in this manner was strangled or broke its neck.

Once, in an effort to catch colies in the winter, a Zulu boy used another variation of snares. The nooses were tied to individual twigs in such a way that the elastic loops would stand erect by themselves with their openings at right angles to the long axis of the sticks, and these noose-covered twigs were then laid along branches of a fruiting *Halleria* tree. Because they fruit in the wintertime when other native fruits are very scarce, these trees attract practically every species of frugivorous birds. The fruits are ideal for this tactic of snaring, because the dark purplish-green fruits, about the size of blueberries and rather insipidly sweet, develop along the branches and twigs rather than terminally among the foliage. To get to the fruit the colies customarily employ rodent-like habits of running along, and in and out, over the branches. The noose-bearing twigs were arranged to take advantage of this habit, and the birds became ensnared.

Snares were sometimes attached to a loose basket woven of vines worked into a crotch of a branch cut for the purpose. Where the vines crossed, these same kinds of nooses were attached with the loop pointing upward. These baskets had apertures about half an inch square so that the contents could easily be seen. After the baskets had been completed, they were filled with fruit and placed where passing frugivorous birds would see them. The birds alighted on the basket, and in crawling over it attempting to reach the fruit, sooner or later pushed their heads through the nooses. Half a dozen birds could be caught at one time in such a trap, and frequently a dozen or more might be taken in one morning. Luckily for the other feeding birds, they would see the fluttering victims, become wary, and would not return for some time. After repeated use in a given locality the trap became virtually useless. The smaller birds that could readily be caught by this means were almost entirely local residents, and it was fortunate that they learned quickly, otherwise they would have been exterminated even sooner than they were. Bulbuls, colies, white-eyes, and a few other birds, especially the collared barbet, were those most frequently caught.

For large birds such as the bush partridge and guinea fowl a fairly large openwork hut of vines was woven. These traps were two to three feet long, and baited with corn scattered around the entrance and along approaches to the trap. Inside was usually a cob of corn tied to a string, which in turn was attached to the trigger. The actual "set" consisted of a little upward-closing doorway that would be jerked shut by a pull at the corn cob. Such traps often would catch the cock and hen and as many as half a dozen young. This was a deadly device, for a time at least, and an effective means of getting meat.

Another method of catching birds was bird lime. It was used in two ways. Dotted over the countryside were large wild fig trees, commonly called *umtombe,* and when one of these trees was in fruit, birds from the surrounding areas congregated to feed. Large varieties of species would soon gather to feast; there were usually representatives of several families, and even orders, present at the same time. Insectivorous birds, even flycatchers, gathered about, for there was always considerable waste ripe fruit knocked to the ground, so that in a few days there would be an abundance of insect food, attracted by the rotting fruit and presumably breeding in it. The native boys would watch for a tree about to ripen, for they foresaw the prospect of roasting meat.

As soon as the birds gathered, boys, preparing for the first of the two methods, came armed with lumps of bird lime (made from the fruit of a mistletoe, *unevu*) and a few dry sticks, climbed into the upper branches of one of these trees, and tied the sticks so that they protruded above the foliage and made attractive perches. Especially in this country of arboreal bird-eating snakes a notable trait of many birds is their habit of selecting bare twigs that can conceal no enemies, and hence make ideal lookouts from which to search the dangerous leafy hiding places below.

When the perches had been properly placed, the boys smeared a covering of bird lime over them and then returned to the ground to wait. The flock of birds which they had frightened away when preparing the traps soon returned, but having seen enemies in their tree, they were even more cautious than usual, and unfailingly would select bare twigs from which they could safely scan the leafy green cover below them. The chief result of this timidity was that up to a dozen birds might be caught at this first return. Like the basket trap, however, this method worked well only in the earlier stages of the game. The birds soon avoided these sticks, and since the bird lime deteriorated in a comparatively short time, the efforts at bird catching were soon abandoned in any one tree.

The other method for trapping with bird lime, used to obtain those species that live on the ground and follow well-defined trails, like quail and francolins, was developed by the Zulus possibly from observing a spider's effective use of multiple strands of comparatively weak threads of silk. When quail and francolins were to be trapped, the natives collected a handful of particularly tough, fibrous grass and smeared the blades with a small amount of bird lime. The blades were then arranged on the sides of a runway so that they would adhere to the feathers of any passing bird, and in this way entangle it. Presumably while attempting to rid itself of the first few blades that had become attached to it, the bird soon became panicked, and in flapping about was ensnared with more and more of the sticky grass blades until like a fly in a spider's web, it became helplessly wrapped in multiple bindings of the sticky material. The method required much lime, and since only a single bird at a time could be caught it was not so commonly employed as the arboreal device.

During the winter season the small boys had more time in

which to hunt birds than during any other part of the year, but even in summer they managed to find plenty of tidbits. It was their regular task to herd cattle and, like small boys anywhere, their ingenuity in escaping set tasks gave them many opportunities to find small game. As a matter of fact, meat was often not as difficult to obtain in summertime as at other seasons because most birds nest in the summer, and it was comparatively easy to find, catch, and kill a luckless incubating mother bird. Most boys waited until fledglings were about ready to leave the nest, when either bird lime or a blow with a stick or a stone from a slingshot would kill both the parent and young.

When the boys managed to catch a coqui francolin, *i'ntswempe* (as big or bigger than our bobwhite quail, and very much like one in color), they often kept these prizes hidden from the eyes of older relatives. The father, who could (and usually still does) claim every cent earned as wages by family members in city work, also unquestioningly claimed his portion of real game, and rewarded the son with only a fragment, possibly the head. Even this grudging generosity resulted only when the game had been brought home voluntarily, but woe to a child so stupid as to be caught trying to keep the prize, or was tattled on by some other jealous youngster. Zulu thrashings were no mere spankings.

Almost as soon as they were put to herding cattle, small boys became adept with the throwing stick. Imps, only six or seven years old, harried birds from bush to bush with their diminutive throwing clubs. Some of the men and older boys sometimes developed extraordinary skill, but on the whole, a great many more sticks and clubs were thrown in getting one victim than would be expected, considering the tameness of so many small birds.

The deadly effect with which some natives used these throwing clubs was often demonstrated by native yard boys when asked to kill one of our half-wild chickens, usually for a Sunday dinner. I recall that one of these "boys" was especially skillful at knocking down the fowl selected to furnish us with meat. He would start a stalk after some particular superannuated hen or surplus rooster and gradually edge it out of the brush into the open. This was often difficult, since most of those chickens were able to fly as well as ring-necked pheasants.

As soon as the fowl found itself out in the open, it would scurry toward the nearest shelter. Then the boy would also step into the open, and with a long seemingly effortless motion of his arm, would send the club whirling along, a foot or so above the ground, so nicely calculating speed, distance, and direction, as infallibly to hit the bird on the head. They were so skillful that I never saw them hit a chicken's body and bruise or even break the skin, and yet their club traveled with sufficient force to kill the fowls outright. Chickens running at what a gunner would call a crossing shot were often decapitated.

While I was hunting one time, a quail flew overhead, and after my miss with a shotgun, a native standing by, threw his club with such force that the bird was cut as cleanly as though with a knife and fell in two separate parts.

The *isifu* is a relatively simple device consisting of a flat, heavy stone supported at one end by a twig resting in the fork of another stick that is inserted in the ground. The twig that supports the stone ends in a piece of tough bark or string usually woven from native fibers with an inch-long twig tied to its lower end in such a way as to make a trigger that allows the stone to fall when anything steps beneath it. The trap is usually baited with fruit, grain, or termites, depending upon the location of the trap and the most probable or the most desired victim.

The *isifu* catches everything that moves onto the twig platform that actuates its trigger mechanism. Of all the small stuff that the *isifu* catches, about the only things that are not acceptable as food to the small African boys are snakes, lizards, and toads. Even these victims are rejected largely because of superstitions and social custom, but as for every other animal that may be caught, a hungry belly serves as adequate sauce; and although most boys will not admit to having eaten mice, there are few indeed that have not at least tried them, and almost all of the rural small fry have enjoyed the larger species of rodents.

In obtaining these tidbits the *isifu* has one drawback—a purely aesthetic reaction to the smashed appearance of its victim. To civilized people with more squeamishness than the African youngster can afford, the usually flattened body of an *isifu* victim looks most unappetizing. In addition, of course, there are the ubiquitous little ants that always seem to reach any source of protein in a very short time. The small boys nonchalantly dust off the ants and proceed to cook their feast.

Today, natives employ the steel trap for catching birds. A small type of this cruel trap is sold widely and used by the natives almost exclusively for trapping birds of all kinds. It is a deadly effective implement for many species, whether arboreal or terrestrial, and the Zulus have rapidly developed techniques of trapping to cover a wide variety of habitats and bird species.

On one occasion a herder boy offered to sell me a female owl that the little barbarian had caught in a steel trap. The bird was almost dead from rough treatment and abuse. She was rumpled and mangled so badly that one leg had been almost severed from her body, yet weak as she was, she managed to stand erect, and with one wing, partially shield her downy owlet. Her eyes were barely open, and as she stood she weaved back and forth, an exaggeration of a common trait, but she was scarcely able to maintain her balance. That she was shielding her owlet was certain, for as it moved she moved also, always so that her wing rested over it. The owlet, with big glaring eyes, hissed and snapped ferociously when anyone approached, but at the last moment always edged up against its mother.

The callous cruelty of native boys makes it necessary for a humane person to stipulate that birds shall never be moved from their nest, nor trapped, nor in any way interfered with if a reward is to be received. But in spite of insistence on these regulations, I was frequently offered birds in such bad shape that the only humane thing to do was to give a small compensation to obtain the specimen, and put it out of its misery. This was also the fate of the owl; the little owlet joined the ranks of its dead ancestors, but, unlike them, its feathery integument is now a permanent museum specimen, of use to human beings for many generations to come.

Large numbers of birds and other game have always been killed by these various devices but in view of the abundance of birds that are still present in the remaining bits of heavy bush, especially where the boys have no other weapons than the "classical" devices, one must assume that a natural balance between fecundity and mortality rates was maintained. Even the air rifle has proved more destructive than all primitive devices evolved by natives, but most deadly of all has been the rapid growth of native populations that deprive the birds of their natural habitats. Today, the arts of traps, bird lime, and clubs have been forgotten by the natives except those living on white men's farms. Deep in the native tribal lands where these customs should survive, few if any boys can demonstrate these skills, for the bush has gone, the sheltering grasses are going, and only barren soil remains. Against the deadliest weapons of man, fecundity and agriculture, scarcely anything but weeds or parasites can survive for many more decades.

6

SMALL-ANIMAL WORLD

The spectacular big-game animals are those that have captured the attention of both biologists and laymen, but there is just as great a reward in the study of small animals inside and outside the game reserves; practically nothing is known of the habits of reptiles and amphibians, not to mention the insects.

In most of the following biological omnibus, place names will hardly be used. In general, each creature occupies a far greater area than the Umzumbe or Amanzimtoti valleys where most of my work was done, and what is said for a species in either of these places until recently was equally applicable to any of hundreds of other similar streams and valleys in Natal. But this situation has been and is changing rapidly. For instance, most of the hillside and streamside bush in the vicinity of Durban has been destroyed and replaced by sugar cane, hence only the cane-adaptable species remain. Elsewhere other forces are at work changing the countryside and its wildlife. These forces will be discussed in the final chapters.

At the present time unspoiled conditions with a fauna that is intact except for large game, can be found only outside the coastal cane belt and the other intensively farmed country. In effect, this means that only the numerous Natal forest reserves, some farms, and a few mission stations, will still harbor most of the small creatures.

If asked to name the most grotesque small mammal of Africa, almost anyone familiar with its fauna might be expected to name the aardvark, earth pig, ant bear (*itsambane*), or more technically, *Orycteropus*. Even the scaly pangolin's oddities fail to equal this Alice-Through-the-Looking-Glass outlandish animal. This largest of all burrowing mammals is rare and supposedly has no economic value; it deserves special mention in the natural history of the country.

At one time its large dwelling holes constituted a serious hazard to a cross-country gallop; they still do, but in a steadily diminishing area. The disappearance of these totally innocuous, very shy, and apparently inedible mammals is probably due almost solely to the spreading agriculture and concomitant disappearance of their only major source of food, namely the above-ground mud nests of termites. A contributing factor might possibly be the use of various anatomical parts of the animals in the native pharmacopeia, although so far as I am aware they have no particularly potent value, at least not sufficient to justify the enormous amount of effort required to locate and dig these animals out of their underground retreats.

Even where the animals are "common," they are more rarely seen than any other mammals in Africa (okapi, bongo antelope, and giant forest pig being possible exceptions), and their passing from the scene is regrettable to all but horsemen. The ant bears, or at least their burrows, were a part of the ecological requirements of at least one plant (a cycad), wart hogs, possibly the striped and spotted hyena, a small black swallow, snakes, and presumably numerous other animals. Before they disappear from all but wildlife reserves, someone should seek to ferret out their role in the control of the termite, and the importance of their long-persisting abandoned burrows as shelters for many of the local fauna, some of which may be so completely dependent upon these dwellings that extermination of the ant bear will affect them also.

In 1953 I saw no burrows in any of the heavily populated native-reserve areas where once they were so common. The native game guards were unanimous in saying that this animal was responsible for the holes used by the wart hogs, and that without the ant-bear burrows the "hogs" were forced to lie up under thick brush during inclement weather where they are miserable and susceptible to cold and wet. The guards base their conclusion on the alacrity with which the wart hogs get out of the rain by running for shelter to their burrow and by their similar haste when cold weather blows into the reserve. This theory seems reasonable, and if confirmed would contribute yet another of those interesting instances of the mutualism in animals.

Wherever farms are situated near the native bush, marauding animals, probably various species of the mongoose tribe, raid the barnyard. When all is quiet, and predaceous animals farthest from the mind, sudden cackling shrieks from the trees harboring the half-wild fowls of an African farmyard announce another raid.

It is extremely difficult to observe these animals at their work. Sometimes a pair of greenish eyes reflects the light of an electric flashlight, or a gray form melts into the black shadows of the bush. It is almost never possible to get a shot at the marauder, and traps in the immediate vicinity of the bush seem either to fail entirely or accidentally to catch some luckless chicken.

(Domestic fowls that survive the dangers of the bush do so by virtue of extreme caution. The more phlegmatic individuals soon vanish. The end result of this natural selection furnishes a breed of hens as harebrained and hysterical as any that ever blessed a barnyard. On one farm an old rooster carried matters so far as to croon his warning cry with hysterical enthusiasm whenever a large butterfly passed close by. As a result, the entire barnyard would be stampeded foolishly a dozen times a day, and even his ample harem finally gave up any attempt to sort the false alarms from the true. This extremity of caution is better than none, however, for in the bush foolhardy hens never reach old age.)

Traps set deep in the bush catch genet cats, *intsimba*, miniature leopards in appearance, and the so-called "weasel," a little, rusty-red mongoose. Also from the bush come the snake weasels, or *inyengelese*. These look like small and very slender skunks, set on legs as unproportioned to body length as those of a dachshund.

Other marauders are almost never caught. Blamed as a raider of the hen roosts is also the Cape otter (*umtini*), supposedly operating from the river and in some localities quite common. Judging by the various signs, however, these African otters seem to confine themselves to a diet of fresh-water crabs and shrimps, varied by an occasional fish or meat of another kind.

Otters are capable of killing fairly large animals, and when hunted by dogs often kill them; but it seems doubtful that otters would ever go far from the rivers on their hunting forays, and still more doubtful that they would be able to climb trees sufficiently well to get the roosting fowls.

It is more probable that the animal doing most of the damage to chickens is the large gray mongoose, the *umhlangala*.

RATS AND SHREWS

In rushes, tall grasses, and in marshy areas, there are cane rats (*amavondwe*), shrews (*ngosongoso*), and various other mammals striving for a livelihood, each filling its small niche. The cane rats have a hard time, for they are the favorite food of the larger species of mongoose; in the past, they also furnished a food supply for leopards, which were once abundant.

Although its natural enemies are being destroyed, the cane rat still fails to become numerous in the old, natural habitats, except in the sugar-cane fields where it has become a dangerous pest. Part of the pressure that keeps down their numbers is doubtless the esteem in which they are held as food by the Zulus. They are so highly valued that although natives frequently kill them, it is almost impossible to purchase specimens. When Zulus do offer to sell these rodents they are usually mutilated. The Zulus share our aversion to eating rats, and the cane rat has an indubitably ratlike tail. However, in a typically infantile fashion, the huntsmen manage to circumvent even their own inmost reactions to rat-eating by an elaborate dodge. When hunting these animals, the Zulus arm themselves with spears and clubs and take to tall grass where the cane rats are known to be present. When the hunt is under way, the men, boys, and dogs form in line and slowly beating through the grass or cover, drive out the game. Here and there men are stationed in likely looking places ahead of the hunt, waiting for a chance to impale or club the animals that have been driven ahead of the line of beaters. When a rat is seen, there is always wild excitement and a great deal of yelling and rushing back and forth through the grass in order to get in the first thrust or blow, which constitutes a brand mark of ownership. The first person to wound an animal in a hunt claims the entire beast once it is obtained, no matter how many others may have participated in the final killing.

After a fortunate hunter has killed one of these beasts, if he follows classical procedures he surreptitiously but shamelessly cuts off its tail close to the body, stands up, and calls to the nearest hunter:

"Oh, look at what I have killed. Is it good to eat?"

And the loyal friend or acquaintance says, "Has it a tail? Let me see. No, it has no tail; it is fit to eat."

Thus, everyone's self-respect is maintained, and the witness to the fact that the animal has no tail is not only assured of cooperation should his turn come, but if unfortunate in that particular hunt, he is fairly certain to profit at the next meal when cane-rat meat is served.

Many species of shrews found along the streams and in the fields are a little larger than our common North American shrews. That is, they are almost as large as the domestic house mouse and are possessed of a voracious and fiery spirit.

When one traps mice in regions where shrews abound, it is common to find that half of the night's catch has been partially or wholly eaten, and many good specimens are damaged beyond salvage. For a time I did not know what animal was guilty of these depredations. Then I baited several traps with fragments of the ruined mice, especially the brains, and caught shrews.

This raises the question as to whether these little beasts would attack uninjured mice or even each other in their hunt for food. Evidently they are not merely ordinary scavengers, even though they would probably not attack an uninjured member of their own species. It is difficult to conceive of cannibalism except in rare instances, but it is quite conceivable that these shrews would attack unwounded mice. Small as they are, they evidently have well-developed carnivorous proclivities rather than the usual purely insectivorous habit. A lion will attack an animal larger than itself, and there is no reason to believe that a shrew, on a much smaller scale, might not do the same thing with equal success.

If the shrews do attack living mice in their hunting forays, they are the most savage carnivorous animal that still persists outside of the big-game sanctuaries, except for the mongoose. A mouse must face innumerable hazards in its usual round of existence, and in the "forests" of reed and grass stems where shrews range in search of prey, they may be one of the major terrors for small rodents. Although not much larger than a man's thumb, shrews are driven by a gnawing need for food, and they have the spirit and strength to find and kill a wide variety of animals.

A little gray shrew, sometimes found in abandoned termite nests, may be as savage in its world as its larger relatives are in theirs, but if so, the Zulus have no inkling of the fact. This tiniest of the local shrews is scarcely as large around as a pencil and about as long as two and a half of the brass bands on a pencil tip. It is a tiny replica of its larger relatives and would furnish barely a mouthful to some of them. During the rainy season these minute shrews must suffer severely, and during this time one finds them frequently on the paths and roadways after a heavy rain lying dead. They do not look as though they had been drowned; they seem to have lain down and died.

It is probably this fact that has given rise to the Zulu belief that of all creatures these are the most timid or that they have been afflicted by some witchcraft or curse. As the story goes, these little shrews wander about in the grass, finding food and rearing their families, but the time comes when they feel restless and start moving off to new fields. It is this restless spirit that leads them to their death, for it has been decreed that to cross the middle of a road or path means death to them. They can walk onto a path or along the side of the path, but once they cross the fatal and imaginary line down the center, they are doomed. It is the unwary or unknowing ones that are found dead early in the morning, so the Zulus say. The most probable explanation for the death of the shrews might be that owls or other predators that have caught the shrews and moved to the open spaces to feed, subsequently examine and reject these animals because of their strong scent glands. Birds are supposed to have a poor sense of taste or smell, but a keen sense would be unnecessary with these shrews. An alternative possibility may be their extremely high metabolic rate and food requirement and hence starvation and exhaustion in inclement weather.

Although the reed beds swarm with life of various kinds and although the shrews hold almost undisputed sway in their own small kingdoms, from outward appearance there seems to be no quieter place in which to live. It is always difficult to visualize the ruthless hunting that takes place in such beautiful surroundings.

FRUIT BATS

Whenever the fig trees, *umtombe* (*Ficus natalensis*), ripen—and somewhere in the bush there is usually a tree that is ripening— the fruit bats with their almost two feet of wingspread gather to feed.

It is seldom that one finds a roosting place of these gregarious animals, but now and again, looking into the dark tangle of vines and branches near the top of a tree, half a dozen to several dozen dark, roughly pyramidal forms can be seen hanging in the sheltered recesses.

Usually these forms are motionless, but even in the daytime, if they are watched for a while, one or the other will be seen grooming or scratching itself, and if there has been the slightest disturbance below, a look through a field glass will show that at least some of the flock will have their eyes open watching for danger.

Curiously enough, when disturbed they often persist in rotating the body from side to side on its long axis. When all else is still and there is no wind, this habit is quite revealing, but, since there is usually a breeze, dead-leaf-brown shapes in a treetop also would be moving if they were genuine leaves, and so by the law of averages for windy weather, safety lies in this restless behavior.

Even when hanging in a treetop, the epauletted fruit bats distinctly show the white tufts of fur on their shoulders and ears, although never as clearly as when the small pocket from which the hair grows is everted. Nothing seems to be known about the function of these white markings, but they are an interesting feature that arouses curiosity.

For the most part, these fruit bats are merely known as swift flying and almost silent silhouettes circling about a fig tree or other tree with ripening fruit. Except for these glimpses, occasional periods of calling, and the scuffling sound when they fight over a cluster of fruit, they are virtually unknown to the residents of the country, both black and white.

I know of only one person to have made a pet of a fruit bat, and her experience should encourage others to follow the example. This young lady * reared a wild fruit bat, "flying fox," from its nursing to independence and had the pleasure of its voluntary nightly return for more than a year. Only toward the end of the year did the usual wild trait of drifting back to nature interfere with observations.

The mother of this bat had been shot, and when she was picked up, a baby bat, almost half the size of the adult, was clinging to one of the nipples. Considerable force, almost enough to injure the young one, was necessary to detach it from the parent's breast, but it took readily to a mixture of milk and sugar or sweetened diluted canned milk, and grew rapidly. Later in life it still enjoyed sweetened milk, and with its very long tongue would lap it from a saucer like a cat or dog. The long tongue is useful in grooming, especially the face, and also the eyes, which remain open even as the tongue passes over them.

Although it has long been generally accepted that the com-

* Miss A. J. Alexander, of Inchanga, Natal. I am indebted to her and her parents for the note that follows.

paratively huge fruit bats (14- to 24-inch wing span) feed almost solely on fruits and nectar, as reported for most of the other species, this individual was tested with other types of food, and it fed avidly on a large green grasshopper, the *umcimbiti,* cockroaches, crickets, and of course flying "ants," that is, termites of the nocturnally flying species.

Most mammals and all known birds, amphibians, and reptiles that feed on insects swallow their prey whole, except possibly the harshest parts but these bats thoroughly chew their insect food, and expectorate the chitinous remains. This was true of even the tender and succulent "flying ants," the sexually developed termites. Tested on ground meat, the bat showed some interest, but finally rejected the proffered food.

Fed on fruit, the bat eagerly ate practically every variety of wild and domesticated species. However, as with most vertebrate animals, even the most favored foods could not be used indefinitely without some monotony-relieving changes; otherwise there was an obvious flagging of interest with final refusal to eat.

The bat was allowed to leave the house at will. It once returned with a ⅜-inch perforation in the wing membrane, and it was seen to pass its tongue completely through this opening during the nightly grooming. The observers feared that this constant licking of the edges of the perforation would prevent healing or at least would keep the edges from growing back together, but the wound healed, leaving its original site practically undetectable. Apparently the licking prevented healing of the wound margin, thus permitting gradual and normal growth to continue until the edge came together and formed a perfect union.

As most bats, these animals rest and feed hanging upside down. This habit introduces many problems that are not encountered by the normal body positions of other animals. While eating fruit, particularly mangoes, which are notorious for the way in which their juice runs and spreads, the pet bat frequently matted its fur with fruit juice. When the bat found such a matted tangle, it lifted its foot and with the parallel claws close together acting as a comb, the foot would fairly vibrate, at blurring speed, until the claws became tangled with dry juice and loosened hairs. It would then clean its "comb" with its mouth and once more proceed until thoroughly cleaned.

A number of interesting observations were made on its manner of feeding. If food dropped from its mouth or from the hand of the person feeding it, the morsel would be caught with the thumb of the wing; the wings were also used for holding edibles, although the large cheek pouches were of particular use in feeding and were crammed with food, one foot assisting in the process. Fruit was seldom eaten on the spot but more usually, and in the open it would seem invariably, was transported from its source to some favorite eating place.

While the bat was eating, juice frequently poured from the mouth, but instead of entering the nostrils, as would necessarily happen in the feeding posture adopted by these bats, the large conspicuous groove running between the nostrils functioned as a trough to carry the juices away and allow them to drip more comfortably onto the ground.

"Personal hygiene" was excellent in the bat. On defecation it reversed the normal position, that is, the body was elevated so that the posterior end was lowest; the same was true for urination, except that on completion, it shivered or shook violently, and in this way scattered the drops anywhere but on its own body.

One of the many notable differences between the fruit bats and the more familiar varieties is their clean body odor, one which has been described as "fruity" or perfumed. It is quite probable that at times these bats feed on nectar, and their long lapping tongues would suggest this. Whenever the Eucalyptus trees are in bloom, it is notable that collected bats have a scent much like that of the Eucalyptus blossoms. Certainly fruit bats always have a pleasant body odor. They have other attractive features such as their large brown eyes, softly furred body, and dog-shaped head; these characteristics save them from being as objectionable to most people as are other species of bats.

The eyes of the fruit bat have not tended to become relatively unused organs, as seems to have happened to other bats, for that near-pet bat could see food at least across the room. Although the fruit bat's sense of smell is good, it was often not used by the pet, which would go through its usual anticipatory motions until the food was brought to it or until it flew to the food. If it had been feeding too long on the kind of food proffered, and was tired of it, it would object and leave it untasted.

During the winter months the bat would fly away, but come back regularly and be fed every night; but as summer approached with warm weather and more food, it was often absent for nights at a time. Gradually, almost imperceptibly, the slim shape of its body seemed to be becoming stouter in the abdominal region. Eventually it was apparent that it was pregnant, and the young lady hoped that soon two bats, mother and child would be returning to feed at the onset of cold weather if it could find the return entrance open to its favorite corner in the warmth of a bedroom.

Before pregnancy the bat had at one time been away for an entire month, but even after this first prolonged outdoor stay of its young life, it hung up in its old familiar corner when it returned.

No matter how long its absence, there was always hope of a return or a double return. This continued to get the young woman and her family out of bed, even at three o'clock in the morning, whenever they heard the wing beats of a large bat passing back and forth across the open transom. By the time I left Africa they were still hoping.

Throughout the north temperate regions, practically all bats are different from the tropical and subtropical fruit-eating kinds, and persons familiar only with the little, insect-eating, half-blind northern species (the "Hallowe'en variety") would have to revise their ideas of this tropical species. Even in more abstruse scientific details the contrast between our better-known insectivorous species and the fruit eaters is of great interest, and this is even true of body temperatures, an important but not widely publicized aspect of natural-history studies.

Most species of our American bats, possibly all of them, are capable of as dramatic changes in their body temperatures as are the cold-blooded animals, the reptiles and amphibians, and with additional studies they will probably be found to out-perform even these fluctuating-temperatured animals. In our American bats, temperatures may drop to near-freezing—they may even have frost on their fur—and they quickly and comfortably relapse into almost complete torpor. From this state they can wake up, shiver, and rapidly generate heat until their nervous and muscular abilities are soon capable of the speedy responses required in flight. Otherwise, their low temperatures also permit them to live for several winter months without feeding, which they could not do if they depended on keeping their bodies warm all the time. Fuel, in the form of food, is essential to maintain heat, and in cold winter weather or on cold nights when insects do not fly, the animals would starve to death without the boon of periodic torpor and lowered energy requirements.

The tropical or subtropical fruit bats cannot resort to this device. They require far larger and more regular supplies of food throughout the year, a possible explanation for their inability to extend their range into far northern or southern regions.

Even the fruit bats can permit a reasonable lowering of body temperatures, but at 20° C. (65° F.) they become restless and begin heat generation and consumption of their own stored energy, or else they must feed in order to prevent their self-destruction. In the wild state, at least from the meager evidence available in South Africa, even on cold rainy days when the temperature of the environment is as low as 18° C. (64.4° F.), their body temperatures are higher than would be expected, for I have collected them under these conditions with the body still as warm as 35° C. (95° F.), not much lower than that of some other mammals.

One reason for a naturalist's surprise at their maintenance of comparatively high temperatures is the enormous expanse of wing membranes, a structure that must have blood if it is to live. But if it has blood circulating through these membranes, then heat would be lost rapidly because the membranes are not only naked and equal to 6 to 11 times the rest of the body surface, but they are so thin that both the outer and inner layers would give off heat whenever the air in their environment was at a lower temperature than that of the body.

When at rest, bat wings have muscle fibrils that contract in such a way that the membranes are thrown into thousands of little folds, partial air pockets that serve to retard heat loss, or heat gain where the surroundings are too hot. Some species at least can regulate the flow of blood in their wing circulation, and in this way help conserve or lose heat, but fur or feathers would seem to serve the purpose far better than this device.

Since fruit bats can be kept as pets, and since they survive in captivity better than most species of North American bats, a study of their lives and physiology would prove interesting as well as a pleasant hobby. However, it is wise that our authorities refuse entry permits to these fruit-eating animals, for if they escaped and became established in the southern states or in California, they could do serious damage.

BY THE MANGROVES

Nature photographers in the United States often find that many of their most pleasant memories are of days spent in a blind set somewhere in a marsh during the spring or summer months. There the gurgling voices of the red-winged blackbirds and the curious calls of mud hens and other larger birds, form a musical background for a setting that is as nearly tropical and primitive as can be found anywhere outside the Tropics proper. In marshes anywhere there seem to be reminders on all sides of those long-past ages when most of the world was tropical for at least part of the year, and when extinct monsters still roamed the earth.

In Natal, as might be expected, days spent in the marshes are as pleasant and stimulating as in the United States. Besides, there is a more impressive reptilian life, and the birds and the many unknown voices and inexplicable stirrings, splashings, and alarm notes, add a sense of tension, excitement, and mystery not to be found in our more domestic scene, not even in the everglades and hammocks of Florida and the Gulf states.

Of the interesting swamps in South Africa, the narrow tongues or islets of the fresh-water mangroves that grow in the estuaries of streams within hearing of the Indian Ocean surf are among the most exotic.

It is always eerie in the dark shade of the mangroves. On wind-less days there is usually a dearth of animal sounds except for those that filter in from the edges, whence come the raucous clattering calls of the giant kingfishers as they fly over open water, or the musical twitter of those reed-bed gems of the same family, the minute azure-and-coral malachite and Natal king-fishers—but these are not denizens of the mangroves themselves. Sometimes in the spring of the year there may be an added note, the squalling of the young of the giant kingfisher, as she feeds them and departs again from the nest in a remote section under high banks bordering the lagoons. At long intervals a pair of hammerhead storks, avian "sad sacks," may drop in for a visit and raise their clangor morning and evening, barely heard at a dis-tance, and from the outer edges of the swamp land may come the melancholy descending and ascending, mellow staccato calls of rain coucals.

In the spring the old mangrove leaves fall from the trees, open-ing the canopy of foliage so that flecks and splotches of sun-light reach the bare soil beneath. When a gust of air passes through the grove, their pattering fall is faintly reminiscent of the autumn leaf showers in the woods of New England states. There is little other to suggest familiar woodlands, for on still, breathless, hot days, in the special silence that is so characteristic of these places, on all sides almost the only sounds are those of individual leaves falling.

When the air is saturated with moisture and the smell of mud, there are also whiffs of the musk turtles' emanations tainting the air. Then, in the close stuffy silence, there is something depressing in the sounds of these leaves tap-tapping their way loudly in this void, until they are silent at last on the mud below.

As the leaves detach and fall they give way almost overnight to the new budding foliage that is swelling, expanding, and spreading a tender green wash of color over the usually somber greens, and each stem and mid-rib of the new leaves is a faint red, so that this new greenness is tinged with a delicate coppery pink.

Along the foreshores the reeds sprout their slender green shafts into the air, and in the reedy pools, frogs and toads maintain a diapason of swelling and falling choruses so that from time to time they drown the more faintly heard calls of the bush birds inhabiting the swamp margins.

For long periods the surface of the deep pools, protected as they are by sheltering trees, will be almost motionless except when a vagrant breeze gently stirs the surface, or when once in a while there may be a loud splash out in the pool where a large fish, monitor lizard, or possibly a python makes some sudden movement.

The bush-covered margins of these swamps are the haunts of the python, most wary of snakes; and of the semiaquatic monitors that can be seen sculling themselves along the surface of the warm water, hunting crabs in the reeds, or simply basking on muddy banks and in the tangles of brush jams left by previous floods.

Before one of my many long sessions in blinds, a local native had warned me of mambas at the edge of the swamp, where other types of vegetation made a tangled almost impenetrable thicket; he was especially emphatic concerning the large and supposedly belligerent black species. As usual, however, even after many days quietly sitting in the blind or prowling the adjacent thickets and mangroves I could see none. Only once did I observe its track as the sole indication of its presence and too-close proximity.

Along the muddy shores of the lagoon and beside the small trickles of water that wound their way through the exposed mud floor within the mangrove thickets, there were tracks of the otter, each footprint as large as a donkey's hoof mark, and there were also tracks of the long-tailed genet cats, and one frequently seen set of unknown footprints, probably those of the large viverrid, the ichneumon or *umhlangala.* Curiously, on neither this nor any other occasion were there any tracks near the blind of the numerous monkeys, *n'kau,* that romped or prowled around, through and over the dry-ground thickets farther from the water's edge, even though their sign was almost always present on the adjacent sand dunes.

In the almost stagnant, debris-filmed pools among the mangroves there were usually schools of fish, floating with the upper part of their snouts out of water, seemingly gasping for air. At a distance they resembled cichilds, possibly of the genus *Tilapia*. By careful watching I could see that they were drawing in a continual stream of surface water and with it the thin floating film of debris. Apparently this habit may combine both respiration and feeding, and in this near-stagnant water they probably profited by a higher incidence of oxygen in the surface layer of water.

To me, sitting in a marsh with so much that resembles conditions of those ancient times when the earth's surface was flat and the climate favored a tropical-like plant life, the presence of the fish and their habit of drawing on the surface film for both sustenance and oxygen suggested the possibility that this could have been a component of the first step toward air-breathing, possibly for lungfish and the more important air breathers, the Rhipidistians, ancestors of the Coelocanths, one line of which lived in shallow water and ultimately occupied the land itself.

The last of the African lungfish are now confined to the lakes near the equator far to the north, and the only known living Coelocanths, descendants of those ancient joint-finned fish that started the long train of evolution of land vertebrates, live rather deep in the Indian Ocean both to the northeast and southeast and, so far as is known, nowhere else.

I watched members of the *Tilapia* genus, the *isikwali* of the natives (*Tilapia natalensis*), in groups of half a dozen or more near the reedy margins by the mangroves. The full-grown females were up to nine or ten inches long and, since they were bass-shaped, amounted to fair-sized fishes. From the isolated position of the pool it was obvious that the fish must have pressed through a veritable forest of stems composed of the various rushes and reeds growing in the water until they had found their way at length into sheltered backwaters. Some were floating in the water allowing the young to play in the comparative safety afforded by the reed barrier. The young are subject to unusual parental care, which probably protects them from most larger animals, although they take their place on the avian bill of fare.

When I came down from the blind and approached one of the

pools, around some of these fish appeared a cloud of what looked like small tadpoles or polliwogs. These were the young, swimming in a loose school, and there were few stragglers wandering by themselves. For minutes at a time the young fish moved about freely in these quiet pools, but when I stirred suddenly the adult fish saw the danger. At such times and with no sign of great alarm, the parent moved a foot or two closer to the cloud of little fish, then for a second remained motionless.

What type of signal was passed to the school of young I could never detect, but without a moment's hesitation there was a concerted movement of the entire school. They huddled closer together and then, looking like a miniature tornado cloud, the small end of the cone touching the mother's mouth, the entire school swirled down as a funnel-shaped mass. In only a few seconds the last of the little fishes disappeared into the sheltering mouth and safety. Because respiration is not hampered to any extent, the young fishes are able to live in the buccal cavity in the region of the gills. The water, entering the parent's mouth, passes through the gill opening—first, however, flowing through and over the mass of young fish. In this way, both the parent and young obtain sufficient oxygen to avoid suffocation.

COLD-BLOODS

With the near approach of summer and the onset of the warm rains in November and December, amphibian life springs into voice almost magically.

At night after a warm rain in every low-lying piece of ground where water gathers—the marshes, the stream edges, the open fields, the groves of banana palms (the locally misnamed "wild banana," the large *Strelitzia*), and even down to the greenery that fringes the edge of the ocean just beyond reach of the highest waves and heavy salt spray—wherever it is moist, the air comes alive with the song of various species of amphibians.

Always at night and often on cloudy days, the chorus of music from these primitive creatures becomes one of the most enjoyable features of the countryside.

Although most of the conspicuously different kinds of amphibians in South Africa have been discovered, those that are only somewhat different, such as those whose mysterious color changes lead to confusion, as well as others that superficially seem almost identical, still remain virtually unknown. The life histories of many species until recently were almost unknown—where they lay their eggs; the conditions under which their eggs hatch and' the young grow to maturity; their enemies; their food; and hundreds of other features of their existence. The little animals are so elusive and usually so well camouflaged that it is difficult to identify each species and associate it correctly with the various musical notes heard from marsh and field. For a time at least a visitor must remain only a listener enjoying an aesthetic experience and a baffling nomenclatural problem.

Unlike most amphibians we know in the United States, many of these animals have extremely novel ways of laying eggs. The curious little toadlike *Breviceps* produce clusters of tapioca-like eggs in burrows above water level, others make froth nests, and the clawed frog *Xenopus,* which spends its entire life in streams or ponds, has intriguing habits and endocrine responses.

When the ground has been thoroughly dampened, and the days of warm sunshine have raised the temperature of the soil, sooner or later there comes a night of torrential rain. Then a drive along the highways is a revelation in the number and diversity of species of amphibians to be seen. These are all tailless kinds; there are no salamanders in Natal.

For only a short time during the early part of a rain, the roads at night become alive with hopping, crawling, or lumbering amphibians, ranging from little mites the size of a bumble bee (some of them true frogs), to larger frogs and toads much resembling those of the northern temperate zone.

As the rain dwindles from its maximum, these animals, all of them dwellers in grass, bush, marsh, or forest, beat a speedy retreat, abandon the highway and other open areas, and seek concealment along the roadside.

Throughout vegetated subtropical countries, the apparent aversion of both amphibians and reptiles to exposure in the open is a notable characteristic. In arid or desert country almost all animals have become adapted to open conditions, and for the most part they can survive only in the presence of wide expanses of sand, bare rock, or grass. The animals of damper areas, however, take advantage of the heavy cover that results from moisture, and seem extremely reluctant to expose themselves to the open where their lack of adaptability to such a markedly different environment renders them vulnerable to attack by the numerous predators that surround them.

To those accustomed to think of Africa as a land of heat, and of South Africa as subtropical and therefore warm, it comes probably as a surprise to learn that the winters, especially at altitudes of only 2,000 feet and higher are often frosty cold. The naturalist who experiences such weather is inclined to be surprised that so many African animals seem adapted to survival under these conditions, and this is particularly so when considering the cold-blooded reptiles and amphibians.

Throughout the high farming country of Natal, it is an interesting experience to dress in heavy woolens, with an overcoat as well, and go for an evening walk in September (early spring). Above, the stars sparkle with winter crispness, the air is sharp and biting, and on a still night the grass is white with frost and crunches underfoot just as it does in the northern hemisphere. Dead cattails at the margins of the pond sparkle coldly, and yet the toads continue calling until late in the night, when it is even colder.

Although tropical and subtropical Africa is supposedly swarming with reptiles, most of them reputedly deadly and aggressive, the facts are quite the contrary. Certainly many are venomous, but there are also large numbers of secretive and timid species, and very small kinds that are not dangerous. Also, the illusion that the dangerous species are numerous leads the visitor to the conclusion that he must be continually alert, and that in the course of his travels he will be frequently engaged in encounters with these animals. If one cobra, one black mamba, one green mamba, and possibly two vipers are discovered by the casual traveler in any six months' period, he should consider himself very fortunate and worthy of self-congratulation on his excellent eyesight.

Many residents of South Africa, adults whose avocation or recreation takes them into the veldt—the broad expanses of uncultivated country—have repeatedly commented on the fact that they had never seen more than one or two species of snakes, even though they had been on the lookout. The inability of most South Africans to distinguish even four or five of the most common species of snakes, including the mamba and cobras, indicates how infrequently these people have had an opportunity to see the reptiles about which they have heard so much, and yet like our equally inexperienced Americans almost every one of them is overflowing with fantastic tales about someone else's experiences with the reptiles.

On the other hand, no matter how poor a naturalist the average Zulu may be, he is usually very competent in distinguishing between several species of the commoner snakes and always seems to know the most venomous ones. This is not so much attributable to fear as to the high esteem in which they hold "medicine" or *muti* made from the bodies of these snakes. Organs of these and other animals comprise an essential part of the native stock of drugs.

One of the most notable features of the reptile life in South Africa is its seeming absence in bush and field. Even though mambas or cobras may be present, and even though one may have traveled through miles of puff-adder terrain, ordinarily not one will have been seen. Yet with increasing knowledge gained from watching rather than immediately killing these reptiles, it is clear that actually many individuals of each species must have been within only a few yards of the hunter. This is because almost all lizards and snakes seek to avoid discovery by man, and they are adept at the art of concealment.

Most African lizards (except desert dwellers) have this exceeding shyness or love of obscurity. There are few species that are as unconcerned about revealing themselves as are some North

American kinds, notably those of the arid West and Southwest. In Natal the powerful Nile monitor and some of the other robust lizards are surprisingly shy, and they are all beautifully patterned from the standpoint of concealing coloration—a protection, supported by well-developed tonic immobility (motionlessness) when confronted by man or other dangers. Normally, when these animals are seen by man, they have inadvertently moved out of their normal habitat, usually driven by hunger, danger from fires, cultivation of the soil, or cutting of their home bush areas.

The near-invisibility of the reptiles and the infrequency with which they are seen in spite of their prevalence, are surprising to a zoölogist used to desert habitats in the United States, where most of the terrain is bare, gravel or rocks, with merely scattering vegetational cover for the inhabitants. In desert regions reptiles can be most effectively collected when one drives along the roads at night, for these animals of vegetationless areas find the open spaces of the roads as comforting and unfrightening as they do the nearby sand or rock habitats, and so lie out in plain sight. They can also be collected by driving or walking over the deserts where the animals sit either in the open or in the scanty shade of the relatively leafless bushes. Even in the near-desert areas of scanty rain, where there is enough moisture to maintain trees, dense grasses, and perennial herbage throughout the year, collecting techniques are different from those used in true deserts.

There are a few notable exceptions to this general rule of inconspicuousness among the reptiles, the most obvious being the so-called blue-headed agamas (blau-skops lizard) with which almost every South African child, black or white, seems familiar, and the active, common, but less conspicuous little diurnal geckos, *isibankwa*, that scramble over the surface of tree trunks or rocky areas yet seem almost invisible because of their color and habits. The agamas are more vividly colored and live on and about the boles of larger trees at the edge of bush-grown areas.

For American naturalists the agamid lizards are interesting for several reasons, for they are the Old World counterpart of our American lizards belonging to the iguana family. The two families show similar adaptations to all kinds of habitats; in fact, a superficial glance could lead to misidentification and confusion between several species of the two families.

Some interesting cases of so-called convergent evolution, really convergent adaptations, can be demonstrated by examples chosen from these families according to habitat and ecological specialization. For instance, the blue-headed agamas could easily be mistaken for some kind of iguana related to our fence or scaly lizards (also known as "swifts"). Both of them have extensive blue markings on a mottled grayish-brown body, both scurry up a tree when frightened, and as they ascend out of danger, they keep to the far side of the trunk; both are insectivorous, and both bob their heads during courtship or when alarmed.

Because of their similarity yet separation in geographic and evolutionary sense, the basically important factor of heat for the two kinds of lizards might be compared.

For all reptiles, in fact for all cold-blooded animals that must depend to a large degree on external heat sources, the problem of the animals' temperature requirements is vital for them and intriguing for the investigator.

The lizards of the American deserts can regulate heat intake, to some degree at least, by the position of their bodies in relation to the incoming sun, as a sunbather on a beach controls sunburn. Many of them are also able to change rapidly from black to almost white. When living under conditions that might otherwise overheat them, they are light, nearly white, with high reflectivity, while at other times, as when cold, they are almost black and thus theoretically at least tend to absorb heat more rapidly.

How much of this color change from black to white, is due to the danger of remaining black on a light surface and thus being conspicuous to ever-hungry enemies, and how much is determined by their need for either heat absorption or protection from overheating remains to be demonstrated. For desert reptiles both needs may be met by the same mechanism. Their capacity for color (albedo) change, and the simultaneous serving of both requirements would procure for them true protective coloration rather than the single advantage of concealing coloration.

I observed no appreciable change in the darkness or lightness of the blue-headed agamas, even though the blue of the head did become intensified while they were in the sunshine and heated to near their upper limit. So far as I could tell, they also failed to orient their bodies effectively to the sunshine when they became too hot, and none of them made the slightest attempt to escape overheating by digging into the ground. Otherwise they seemed to be much the same as their iguanid American counterparts, since they started panting at a body temperature of 38° C. (100.4° F.) as do some of our North American species, and they were obviously suffering and not far from heat death when their body temperatures rose to 42° C. (107.6° F.). They were also practically identical with some of our larger scaly lizards, which they approximated in ability to absorb heat at a truly remarkable rate. In several of my experiments with agamas, each time on comparatively cool days when the sun was shining but the air temperatures were between 18° C. (64.4° F.) and 22° C. (71.6° F.), the lizards would have their temperature increased above that of the surrounding air by as much as 18° C. (64.4° F) in the amazingly short time of ten minutes. For a cold-blooded animal, one incapable of regulating its own heat even to sustain a temperature of 20° C. (68° F.) above that of the air blowing around it, this is sufficient evidence of their adaptability.

Whether their differing reaction toward overheating is a result of their evolutionary history or simply a matter of special adaptation to an arboreal life may never be known, but the manner in which their behavior seems to fit present needs is logical to our way of thinking. Under natural (free) conditions, when they are overheated they need only pass around the trunk of a tree into a shade that is always available to them a few inches distant. That they have no impulse to dig, in order to get out of the heat, seems reasonable enough, since shade temperatures drop rather rapidly within the branches of a tree, and the lizards are excellently adapted for climbing but seem to be rather mediocre diggers. If color change is chiefly a matter of heat regulation, as has been intimated, it could logically be expected that these animals would be benefited by changing their color or shade from

dark to light, as needed. It seems more likely, however, that concealment is their chief requisite and, if so, their already admirable
matching of the bark of trees could only be adversely affected by
a temperature-induced color change. As to the seemingly conspicuous bright-blue head of the adult animals, especially the
males, this remains as enigma.

The head-bobbing of these and many other lizards (and birds),
including the rapidly repeated "push-ups" performed by many
species of the iguanids, may be necessary because of the lateral
position of the eyes, which probably limits their depth perception. Some degree of depth perception might be achieved through
bobbing the head. Since the reaction usually appears at dangerous or stressful moments, as when males are approaching combat,
or when the lizards or birds are nervous or exposed to danger, one
might guess that the need for binocular vision accounts for the
"push-ups," rather than, as usually suggested, for sex identifica-
tion, intimidation, or species recognition. It might be possible to
conduct experiments to see if some genuine visual advantages
accrue through head movements.

Although the natives are well aware that these lizards are not
venomous, they fear them greatly, and nothing can induce a Zulu
to catch the lizards or even assist in their capture. Why they fear
them so greatly seems to be unknown. If this is the species supposed to have been sent to the world to bring the message of
death, then certainly the agamas are as greatly feared for their
part in the legend as the chameleons are hated for loitering along
the way when sent in the race to bring the opposite message, that
of eternal life.

Among all lizards of the world there are probably few that can
compete with true chameleons in changeability of color, bizarre
ocular movements, methods of capturing their food, and amusing
locomotion. They are also beautiful in their arboreal adaptations.
Their habitat among the cool foliage of trees and bushes is one
of the most delightful and should stir to poetry the pen of some
perceptive and imaginative writer.

The chameleon (*Chamaeleo dilepis*) is par excellence a dweller
in a temperate climate, and definitely seems to avoid heat and

sunshine as soon as its body temperature exceeds about 33° C. (91.4° F.). It also has remarkable ability to sustain nightly exposures to temperatures close to freezing.

Pet chameleons are amusing in behavior and color changes. They are so different from all other reptiles, that in many respects they take the place in human affection or indulgence in South Africa which the horned lizard has won in the western United States. They are so easily recognized that even children soon learn that *these* reptiles at least are harmless. Children will soon advance to the stage of cautiously touching them or eventually keeping them as pets. Their odd ways serve to make of them the most favorable South African introduction to the Reptilia.

In spite of this widespread familiarity, many people still cling to the old belief that these lizards can change color at will, to match any coloring in their background. Although the pets contradict this notion every day, it seems never to penetrate the average mind that they do not perform traditionally. Strangely enough, even people who have kept many of them as pets seem to feel that their own individuals must have "let them down" by not conforming to the old legend.

Actually the chameleon's rapid color changes constitute a baffling problem for the interested observer. Of three healthy pets that I kept under observation for many weeks, for a time all of them in one box, one was usually in the dark-blackish color phase while asleep at night, another was a chartreuse green, and the third and largest was usually bright pea green. However, this was not invariable, for sometimes two were chartreuse and the third was bright green. Under emotional stress the most marked changes took place. One of the best means of causing stress was to place a small snake on the same branch with the lizard. Two of these particular pets tended promptly to change to canary yellow or nearly chartreuse on the yellowish side, while the nonconformist third one was likely to break out in the most surprising pattern of black spots and then change to yellow. But at another time, all turned dark. In daytime, on sunny days when they proceeded calmly about their business, they remained rather consistently green whether perched on brown branches or leafy boughs. When I placed a snake among them, no matter what kind, there was a most amazing exhibition of neuromuscular adaptation consisting of branch shaking. Each animal exerted rapid lateral movements that shook the leaves as though a wind were blowing, but since they did this when there was no wind, and since the leaves would move anyway if there were a wind, it was difficult to conceive how they got any advantage from this curious habit.

A more practical expedient was that of one of the three which, when thoroughly "convinced" that there was danger present in the form of a snake, repeatedly and deliberately let go of his perch to fall headlong to the ground. When I put him back, the performance was repeated again and again, even after he landed on his head. On the ground the celerity of locomotion was entirely at variance with the usual halting and vacillating pace of chameleons. It is curious that only one of the three behaved in this manner, while the others either froze to motionlessness (by far the best tactic, a mere human would believe) or indulged in branch-shaking. So far as comportment was concerned, they certainly exhibited individualism, more than one would expect of reptiles, especially of a species so well adapted to a special environment.

I saw one of the best exhibitions of nonconformity when I removed all three from a cool room at 20° C. (68° F.) in the early morning and placed them on a leafy branch in the sunshine. One promptly turned almost black, one remained bright green, and the third a canary or chartreuse yellow. There appeared to be no reason for these diverse responses except possibly a different chemical and hormonal make-up. However, one of these individuals dominated the other two, and I later found that this dominance undoubtedly induced fear in the others and was thus responsible for their odd color changes.

On one occasion when I had several chameleons available, I experimented again with a snake. I allowed a bird-eating snake, *Theltornis kirtlandii*, to crawl into their vicinity, and although it was too small to have been able to swallow any of the three lizards, its presence immediately evoked strong reactions. One chameleon remained basically green but broke out into a beautiful black polka-dot pattern, after which it became chartreuse, and the others soon followed suit. All then started the leaf-shaking tactic. On examining these animals on their shady side it became evident that for some reason, while one remained largely green, the others were uniformly colored except that one had the right rear foot green, the other was flecked with green only on the shaded side.

Just how light alone could have brought about this difference between the illuminated and dark side is difficult to understand, but if temperature of the surface was the effective factor, then some thermally responsive process could have reacted locally within the skin according to the heat-absorption pattern, and in the cooler, shaded areas the chemical or physical change might have been suppressed or retarded.

Harmless species of snakes were equally obnoxious to the chameleons, and it was immaterial that the *Theltornis* used in this experiment was one of the back-fanged species, thus poisonous to a degree.

THE MONITOR

Among the animals that frequent the neighborhood of streams and rivers and make the dense growth of these localities their habitations are many reptiles. Some of them are poisonous snakes of certain species, and some are harmless, as, for instance, the big and blundering Nile monitor lizard. This reptile often seems to have a special penchant for crashing through the undergrowth at the most inopportune times, usually just as one becomes engrossed in stealthy hunting.

This large and interesting reptile, known variously as leguan, lagavaan, iguana, xamu, or qamu (*Varanus niloticus*), is the object of hostility by animal and man. Of late it has come to be fairly well known through its widely distributed integument in the shape of purses or slippers.

Among the Nile monitor's relatives are some of the largest lizards remaining on earth today, the immense lizards of the

Komodo Islands and the smaller Australian, Indian, and Malayan species. Although the Nile monitor is not so large as some other members of its family (Varanidae) it attains an imposing length; there are some reports of specimens six feet long, and it has been popularly reported as long as seven feet.

The monitor is hated because of its frequent raids on hen's nests and young chicks, and since one monitor is capable of consuming eight or more eggs at a time, and since they bolt the eggs whole, it takes only a few minutes to do considerable damage. While raiding for young chicks they will sometimes boldly come up to a house, and unless there is an exceptionally plucky dog around they usually manage to escape.

It takes a courageous dog to attack and kill one of these animals, for they are fierce fighters when cornered, and the larger ones can give a good account of themselves in a rough-and-tumble. They are equally at home on land or in water, and their tail, which is especially adapted to locomotion in water, happens through this same modification to be an excellent weapon. The monitor's tail is somewhat flattened laterally like that of an alligator or crocodile, but it is triangular in cross section near the tip and exceedingly muscular toward the base. This construction makes a good sculling oar, and the sharp, hard edges or corners make it an effective organ for dealing painful blows.

When attacked by man or dogs the monitor uses this tail with great effect for whip-like strokes right or left with equal facility. The natives claim that one blow from this weapon will often blind a dog permanently. As they put it, the tip of the tail pops the eyeball. On one occasion I attempted to catch a large specimen alive and received the full force of a blow on my arm. For several days the skin was slightly discolored and looked as though it had been struck by a whiplash. I was convinced that one could quite easily be blinded by such a blow.

After a dog has rushed into close quarters and escaped damage from the tail, the battle is by no means over. At close range the monitor resorts to teeth and claws, and although the teeth are neither very long nor very sharp, they are backed by such powerful jaw muscles that it is a common sight to see a rather savage dog back out of a fight, holding its head on one side to ease a crushed and mangled ear. During the in-fighting, the monitor seizes its victim by the head, rears up, places its long claws and powerful legs so that they can exert a downward pull on the victim's back, then with head and forelegs strained to the utmost, there is ample power to break or dislocate the backbone of a smaller animal.

Although a few people seem to think that the monitor is venomous, its bite is innocuous except for possible infection, and it is harmless except for its power and swiftness. It will fight only when cornered. Ordinarily it is so shy and elusive that few are caught without traps.

The life history of the monitor, big though it is and often seen, for a long time escaped investigation by many curiosity-minded persons. Up to 1925 it was supposed to nest in only one way, the conventional method followed by most members of its family, namely, by deposition of its eggs in sandy soil close to a river. Actually the Nile monitor resorts to an ingenious innovation in egg laying, one which, in Natal at least, seems to be general.

It is strange that even Zulus were, and mostly still are, unaware of its laying habit. Possibly this unawareness may have aided in keeping the details of the monitor's life history a closed book despite the fact that it often lays its eggs in close proximity to, and in plain sight of, human habitations.

The Nile monitors inhabiting Natal deposit their eggs in termite nests. Wherever these monitors are found, adjacent hillsides are dotted with the large, rounded nests of termites (*Nasutitermes trinerviformis*). These are composed of hard, clay-like soil baked by the sun so that only by the use of a shovel can one break through the hardest of them. Throughout the dry (winter) season when the surfaces contain very little moisture, the nests are resistant to damage. Because they are about two feet in thickness and proof against invasion by the voracious little meat-eating ants and other potential egg eaters, they are unusually secure hiding places for such precious objects as eggs, and the Nile monitors take advantage of this protection and lay their eggs in the center of the termitaria.

Supposedly about February or March, when the rains have softened the outer layers of the nests, and the monitors have enjoyed a nutritious and abundant supply of food for about three or four months, conditions seem to be right for egg-laying. During the daytime the female lizard digs her way to the center of the termite nest, backs out, turns around, and backs in again. When all eggs have been laid, she partly covers them, but not thoroughly; that is unnecessary, for as soon as a termite nest has been broken into, thousands of the little termites swarm down to the site of the break and start repairs—almost before laying has been finished. But they are not able to damage the tough skin of the mother, so the eggs are deposited in comfort.

Within about an hour the eggs are covered by a thin layer of clay, the material used for the new cells of the nest. As reconstruction progresses, eggs are simply incorporated into the nest. A few hours later only the damp earth over the site of the break shows that an unusual event has taken place, and in a few days even fresh clay of repair has dried out and become the color of the rest of the nest.

The eggs remain in the warm interior of the termitaria from this time to September or later, probably for about nine months. It is difficult to imagine a safer place for otherwise unguarded eggs. The inside of the termite nest retains the same humidity and almost the same temperature for month after month. Even after days of heavy rain, or frosty nights, the eggs are still quite warm and snug. Pouring rain fails to penetrate to where they are lying, and the thousands of small chambers and cells of the nest insulate them from the cold. In fact when a cold thermometer is thrust into the nest it becomes wet from condensation. Even after weeks of cold dull days, ($10°$–$16°$ C. or $50°$–$60°$ F.) I never found temperatures in the eggs to be lower than $25°$ C. ($77°$ F.) nor even after long hot spells were they warmer than $27°$ C. ($81°$ F.) and the air within the nest was invariably highly humid.

At the end of the incubation time, in the spring months (September through October), the young begin to hatch. First one egg splits, then another, and as each egg opens, the excess liquid from within flows into the clay material of the cells that imprison the eggs and young monitors. If it were not for the softening effect of this liquid, it is doubtful whether the monitors would ever be able to escape from their shelter. Neither parent returns to the nest nor shows any interest in their progeny. The young need no attention, and yet, except for this little circumstance of the excess moisture, every single one of the brood would probably remain imprisoned in the center of the termite nest and die within a short time.

As the clay cells soften, the struggles of the young break down the surrounding walls, and soon an infant monitor enlarges the space into which he first moved. At the same time, on all sides of him others are doing the same thing. There are from a half dozen to, in one instance, thirty-six, all in a central chamber. As each hatches, the moisture increases, and in a few days the center of the nest is converted to a large flask-shaped cave, completely closed from the outside. For a few days more, the young monitors remain in this warm, safe refuge, continually engaged in working a passage by which they can escape.

Instead of digging a horizontal exit toward the side, seemingly the simplest and easiest direction, they burrow a vertical passage. The reason might be that the occupants of a termite nest placed by chance in a hillside could never escape if they dug in the direction of the hill. From the top of the flask-shaped cave they dig a narrow passage through which they can pass only one at a time. They do not all leave at once, and often a single individual may poke its head out of the top of the termite nest and, after looking about, withdraw and pass back down into the center.

In the darkness of their retreat they live peaceably, but as soon as their nest is opened and they are exposed to bright sunlight, they begin to fight one another, and in less than a minute all that can be seen is a writhing mass of savagely struggling little monitors. Although only nine inches long when hatched, they fight with the same determination and tactics as the adults. The best way in which a free-for-all fight can be stopped and the combatants separated is to put them into the dark again.

The described mode of egg-laying does not apply to Nile monitors in the more northern regions of Africa. Somewhere in Africa there should exist a line of sharp demarcation between the southern forms, laying in termite nests and the northern kind, requiring sand. There could scarcely be any typical subspecific intergradation between populations, for the mode of egg-laying seems to be a sharply incised physiological requirement.

The greatest dangers to which the embryonic monitors are exposed are excavations by the ant bears, and trampling of nests by domestic bulls. The aardvarks dig into the side of a termite nest, protrude their sticky tongues, and so manage to collect a meal of termites. Often in digging into the nest they do not go very deeply so that probably a good many clutches of monitor eggs might narrowly escape damage even where ant bears are common. The natives sometimes construct floors of clay material from these termite nests, and yet, strange as it may seem, I found no native who had ever seen monitor eggs or even heard of them in the termitaria.

The young monitors soon desert their safe refuge and plunge into dangers that confront them on all sides. Although they are fighters from the time they push their snouts from the shelter of the egg, in a fight they are handicapped by their small size.

When they first leave the termite nest, the muddy soup of clay and moisture covers them from nose to tip of tail. Consequently when they scramble down the sides of the termite nest they almost match its color, and on the ground, on the way to the nearest stream (never far from their hatching place), grass and coarse brush only slowly remove this protective layer which aids them in matching the background of their new environment.

When they reach the stream, where the lights and shadows are more intense on water and sand, they plunge into their new element, perfectly at home. After only a few minutes of activity, swimming and crawling through the damp weeds and grasses of their new habitat, the layer of clay washes off and exposes their real color. Like their parents just after skin shedding, the young are black, with bright, wavy bands of yellow. In this new environment, where the sunlight is reflected so brightly by water and wet sand or mud, and where by contrast the shadows are stronger, the young again blend perfectly.

The protection that comes from such a color scheme, both before and after taking their first bath, must be of great value to
them. Birds and predatory animals hunt them in the fields and
along the streams, so that despite the numbers hatched each year,
and despite their concealing coloration, only a few grow to maturity. At hatching time young monitors can always be found
along the stream banks. A few weeks later only an occasional
small individual can be seen. A few months later the young are
even more scarce, but they will have almost doubled their original size. A year later the yearlings are scarcely more common
than adults. Enemies have taken their toll, and only the hardiest,
wariest, and luckiest individuals are left to replace the losses
among adults.

Although the termites in whose nest the young monitors remain for several days are succulent food, they are ignored by
young lizards. It would seem quite logical that for the first few
days of active life the termites would be used as a source of
nourishment—they are present by the thousand, drowning in
the mud of the hatching chamber; seemingly they would provide
an ideal source of food, but there is no evidence of their being
eaten. The yolk material of the egg from which the lizards
hatched, at this stage of development enclosed in the monitor's
body, must suffice as a source of energy until the monitors are on
their way to a stream.

Associating with the monitors in the aquatic habitat is the interesting little turtle *ufudu* (*Pelomedusa galeata*). The natives
are familiar with its eggs, which are laid in the sand along stream
and river banks, and many assert that they are the eggs of the
Nile monitor. The eggs are pure white and, unlike those of the
monitors, possess a hard shell, which would in all probability
greatly resist attacks by invertebrate carnivores and so remove
at least one hazard of this less perfect environment.

Only one other reptile in this region lays hard-shelled eggs, a
little lizard, *isibankwa* (*Lygodactylus capensis*), commonly
found in stone piles, around old buildings or running rapidly over
logs or other places affording shelter. They do not appear to be as

numerous in similar localities in the bush, and evidently prefer the hot, dry exposures in more open places.

When their eggs are first laid, they are white with a faint pink tinge due to transmission of light through the egg contents, but as development proceeds, they become darker, lose their pink tint, and eventually, just before hatching, show darker patches where the body of the embryo shows through the shell.

These curious little eggs, about the size of very small beans or the seeds of the Kaffir-boom tree, are often yoked together in pairs or threes, attached end to end or at a tangent to their long axis. They are laid with apparent carelessness, since they may sometimes be found unhidden, dropped into mere crevices of a rock or brick wall, occasionally placed on the ground, or, more rarely, laid in the open. Most frequently, they are found under flakes of bark or stone, fairly well hidden, and so located as to receive ample heat.

A collection of these eggs placed in a bottle with an abundance of air, and kept in a warm place, away from direct sunlight, will usually do quite well, and most of the eggs eventually will hatch. In this respect they differ markedly from most reptilian eggs which are notoriously hard to maintain under the precise conditions needed for hatching. The resistance to drying and the hard shell are undoubtedly the reasons that permit carelessness in laying and thus make the eggs so easy to find.

Even some arboreal chameleons resort to the ground to lay their eggs, and deposit them in a cluster two or three inches underground, where they will escape the perils from animals as well as great fluctuations in temperature or humidity, an important factor in the development of most reptile eggs.

The numbers of snakes are dwindling in many areas. Yet, they still appear from time to time, and almost all of them, nonvenomous or venomous, inspire fear among the natives. Sometimes the fear is justified, but the natives are frequently found laboring under a misapprehension since most snakes (even in Africa) are harmless. Some snakes figure in superstitions, and among these are a few harmless species. Indirectly all snakes enter into the life of the native either through religious beliefs, superstitions, "medicine" (*muti*), or because of their dangerous nature. It is sometimes difficult to draw a line between superstition and religious belief.

One of the most dreaded reptiles is the entirely harmless file snake, *Simocephalus capensis,* an illustration of the manner in which a harmless variety may become an object of dread.

According to the natives, death always follows the visit of the file snake, and if one is found crawling about in dwelling enclosures, nothing can convince the natives that it is not a prophecy of disaster. Because no time limit is specified, and since the death rate among children is very high, it is easy to understand how such a prophecy may be fulfilled with sufficient regularity to confirm the natives in their belief. While questioning one of the Zulus regarding a specimen that had been killed, I was repeatedly informed that the snake was so poisonous that by touching it I was endangering my life. To the Zulu the fact that no harm resulted was merely, as always, an indication of the invulnerability of the white man, and the demonstration could not be used to convince the natives of the snake's harmlessness.

Another snake which the Zulus fear greatly is the rock python (*Python sebae*). It is possible to find a few natives who state that it is poisonous, but for the most part, their dread of the reptile is inspired by its size and strength.

Ordinarily I am inclined to assume that where there is smoke there must be fire, and yet I was unable to obtain any authentic reports of an unprovoked attack by a python. Provoked attacks by a cornered reptile might easily occur, and in this way give rise to a conviction that the animal is dangerous. It has sometimes been argued that a fairly general belief in the legend of a python's dangerous nature must constitute at least suggestive evidence that at times it may attack without provocation. Nevertheless, in the light of some of the other beliefs, notably that of the "dangerous" nature of the file snake, it is impossible to have much confidence in the theory.

Although the white settlers are inclined to believe native legends and add many stories of their own, they profit only in a strongly negative sense by scaring themselves to the point of panic when it is known that one of the larger snakes is in the vicinity. That it heightens the drama of living in South Africa cannot be denied. Because dangerous large animals are a thing of the past, it is now chiefly snake folklore that gives the visitor to the South African subtropics the thrill of adventure that the stories are intended to provide. Without some kind of danger to talk about, the hero returning to his homeland would be sadly at a loss for stories of hardship and adventure. By the stay-at-home, every traveler to Africa is *ipso facto* supposed to be a man inured to danger and hardship. Who would be willing to blight this flattering belief?

Here as elsewhere the imaginative narrator cannot be persuaded that *his* snake encounters have not lived up to the best. As a matter of fact, when a student of snakes attempts to explain facts to these victims of their own stories, he is met with either refusal to believe the facts, or with a flare of temper because the teller of tall tales does not want to be humiliated or to retract all the stories with which he has regaled his friends for years. To American herpetologists this is not an unfamiliar reaction. There is no difference between tale-tellers in Natal and their many counterparts in the United States except that South African snakes are possibly the more imposing.

In South Africa, as elsewhere, the snake folklore and legends of both black and white men dramatize only the danger from snakes, and the favorite stories characterize the African mambas, especially the black mamba, with habitual, unprovoked aggressiveness and wanton attack. Actually, most observations strongly suggest that the facts are the opposite of popular lore. As a matter of experience among herpetologists, these snakes ordinarily seem to be as timid and self-effacing as the better-known and harmless snakes of the northern hemisphere.

Both the highly arboreal green mamba, *Dendraspis viridis* and the black (or brown) mamba, *Dendraspis polylepis* even today are not uncommon in a few localities. Of the two, the black species is most greatly feared.

It may be admitted that the black mamba may occasionally attack without a known reason, that is, without recognizable provocation. However, my personal experience and analysis of all stories that I investigated seem to indicate that from the snake's point of view, at least, there was provocation. There are, of course, individual snakes of all species that behave in wholly unexpected ways, and this is certainly as applicable to the mamba as to a harmless gopher snake in the United States. Aggression by the mamba, however, usually results in the death of the person attacked. Aggressive snakes are very rare exceptions to the rule; they gain nothing by aggressiveness and in fact may endanger their lives, no matter how successful the attack. Snakes are fragile and easily killed, hence their extreme timidity. Even their occasional defiance of man merely camouflages fear.

Snake hunting would be easier if more reptiles were prone to stay and fight rather than make haste to the safety of impenetrable thickets. After my experience of two "attacks" by black mambas (one 10 feet long) and of escape unscathed, I feel I can state that these snakes certainly are not the superhuman demons of speed and continuing aggressiveness that stories would lead one to expect. In both instances the snakes were satisfied with a single "pass" at the human aggressor, which made me withdraw momentarily. They did not attempt to follow up by pressing the attack; on the contrary, they tried to escape to cover. In all my other encounters with the black mamba, the snakes either made every effort to escape, or they remained completely motionless in the conventional, concealing pattern of behavior of even harmless snakes when surprised but not threatened.

The black mamba is longer, heavier, and more terrestial than the green, but it is certainly not confined to the ground, for it is often found fifteen to twenty feet up in a tree; however, it is usually seen among the lower branches and tangles of vines, where its color renders it more nearly like the branches on which it crawls. As for the other terrestrial snakes, the numerous underground cavities made by the larger species of termites provide them with excellent shelters from both enemies and climate. These cavernous underground termitaria with their openings above ground, especially in those that have been abandoned, form favorite residences for many kinds of snakes. My own experience in Natal indicates that bee hives are quite as often found underground in these abandoned termitaria as they are in trees, and it is probably the denning of the black mambas in these abandoned termitaria that has led the natives to conclude that the mamba also likes honey and that this is the reason for the not infrequently encountered combination of mambas and swarms of bees.

Intimidation is a fairly common expedient employed by snakes, but it is mostly confined to bluffing. The frightening devices are often limited to behavior patterns such as in the simulated strikes by small harmless snakes that successfully "trade" on the widespread fear of any ophidians. Probably the deadly effects of bites by the few venomous species have saved fully as many lives of their feeble and harmless brethren as of their own. Some species of snakes have such devices as the elongated ribs that form the support for the hoods of the hooded cobras, color patterns that simulate those of dangerous species, hissing, inflation of the body to give the appearance of larger size, feigning of death (which is supposed to possess some advantage), or inflation of the throat, revealing new and entirely different patterns.

One of the most surprising of these displays is found in the bird-eating snake, *ukakoti* (*Theltornis*), which so inflates its black-and-white marked throat and cephalic body section that it gives the appearance of a snake several times its real size and presents a most fearsome appearance, quite out of keeping with its normal near-perfect camouflage and inconspicuousness. The throat-swelling device is employed only after the snake has been captured. When this sudden flashing expansion of the anterior regions is combined with a simulated or a real strike, a man will be inclined to drop the animal, which then will promptly make its escape.

The *Theltornis* is one of the most difficult snakes to see. It is vinelike in body proportions and color, and it has neuromuscular coördination that leads it to assume vinelike configuration and immobility for long periods. The combination of habits and habitat make it almost impossible to see in the first place, and even when one has seen the animal, it is extremely hard to find it again if one looks elsewhere for a moment. The task of relocating an animal after it has been seen and pictured in the mind is testimony to the success of its camouflage. *Theltornis* is usually seen by accident and not as a result of search. Although called "bird-eating snake," examination of its stomach contents provides convincing evidence that it should be called "lizard-eating snake."

With the notable and well-known exception of the *Dispholidus* (or boomslang), the back-fanged snakes have been considered relatively innocuous. Although it has been suspected that their venom might be fairly toxic, the position of the feeble fangs far back in the mouth was believed to render them nondangerous because repeated imbedding of the teeth in the victim seemed necessary to inject sufficient venom to be effective. However, the recent illness of D. G. Broadley, herpetologist of the National Museum of Southern Rhodesia, shortly after a bite by *Theltornis* indicates the opposite, and it also serves as a warning that other species may be lethal.

Another case was the death of F. J. de R. Lock, a Tanganyika game ranger, following the bite of one of these snakes. Autopsy findings indicated that the venom is highly haemolytic, and in sufficient quantities will soon cause widespread and fatal damage. These experiences conflict with the experience of Dr. Pringle, director of the Natal Museum, who tells me that he has seen six of these snakes simultaneously bite one of his assistants, who thereafter showed no ill effects whatever. My own experience is confined to an incident that happened when I removed a *Theltornis* after the experiment with the chameleons reported earlier in this chapter. I was about to take pictures of this snake, and although aware that reasonable caution should be exercised, I was not alarmed at the threatening behavior of the little creature and felt confident that if it did bite I would have enough time to disengage it before it could work its fangs into position to imbed them. This is a common belief among herpetologists. To my great surprise and slight consternation the snake struck with its gape so widely open that its fangs came to bear instantly, no preliminary chewing needed. One of the tiny teeth punctured my ring finger near its middle joint, the other fang missed. Following the bite, the minute wound bled freely and persistently, enough so that it seemed impossible to believe that any venom could remain in the slight puncture, and when slight nausea, headaches, and a faint feeling of muscular tremors followed and persisted for some hours, I ascribed the symptoms to psychosomatic causes and continued my work until it became more comfortable to lie down. The symptoms were annoying to my self-esteem because I ascribed them to fear, but since I still believed the snake to be comparatively harmless, I was puzzled why I should react in this manner. It was only after the de R. Lock incident, happening later, that I admitted the snake could inflict a possibly dangerous, even fatal bite.

Probably the most significant part of the experience is the discovery that at least this species of back-fanged snake can strike and implant venom with practically the same celerity as more dangerous snakes and that its undoubtedly lesser danger to man or large animals may be due to the smaller size and inefficiency of the fangs.

In Natal the only dangerous snakes that are apt to be seen with any frequency are members of the family of true or Old World vipers, and of these one of the most interesting is the notorious puff adder, *Bitis arietans,* or *i'bululu* of the Zulus.

Part of the interest arises from the fact that these animals are, to some degree at least, ecological counterparts of the rattlesnakes of the New World, the pit vipers, and there is some superficial similarity between the two groups. The puff adders are ground dwellers, prey on rodents, and are also capable of producing a distinctly audible sound. It is a prolonged, slightly resonant hiss, not so loud as the castanet-like sound produced by "rattlers" and sometimes obscured by the noise made by one's own walking through tall grass or weeds, or even by the sound of wind blowing past one's ears, but it is often sufficiently loud to warn human beings or domestic animals. The methods of locomotion employed by the vipers include those ordinarily seen in rattlesnakes. There are differences in coiling and striking postures; also, as in our more familiar snakes, there is a great difference in color among the puff adders, apparently depending on the hue of the soil on which they live. This color-matching is probably more notable in the rattlesnakes of our southwestern United States. The nonblack parts of the patterns often closely resemble the soil, being reddish from red-soil areas, and with the yellows also varying to match the soil.

There is as much popular misinformation about puff adders as there is about rattlesnakes, and some of the accounts have notable similarities. Both of these widely renowned species are supposed to be savage, aggressive, and agile, capable of jumping high off the ground and, in the case of the puff adder, jumping as high as a man's thighs or shoulder. Just from the appearance of the snakes this seems most unlikely, because they are heavy-bodied even by comparison with rattlesnakes, which are by no means lightweights, and their locomotion is slower than that of even an adipose rattler. The fastest progress demonstrated by a thoroughly warm puff adder was at a rate of only 750 feet per hour as compared with a calculated maximum rate for even slow rattlesnakes of 5,000 to 10,000 feet in the same time. Actually, neither snake can sustain these high rates for more than a few seconds, possibly a minute, and there are probably no land mammals that would not be capable of escaping direct pursuit by either snake.

When the puff adder is on the defensive, poised to strike, the picture is quite unlike that of their pit-viper counterparts, who usually elevate the forepart of the body and carry it in more or less S-shaped form well off the ground. The puff adder also employs a similar S-shaped flexure, for it is only by means of sudden straightening out these segments that the snakes can strike, but it carries the head close to the ground and remains far more immobile than most rattlers.

Even when it crawls toward shelter this low posture is held; the snakes do not elevate the forepart of the body while they search for a hiding place, and when they come to a pile of dead leaves or grass the head is wedged under the debris, the body pushed forward as the snake insinuates itself under shelter that would seem incapable of concealing an animal a tenth its size. The ability of puff adders to hide is uncanny. I saw one that was two and a half feet long and as thick as a man's wrist successfully creeping under (and coiling its full length out of sight) a thin pile of debris about the size of a baseball cap. Even after a careful scrutiny I failed to see a recognizable part of it.

It is undoubtedly this phenomenal capacity for concealment and the snake's invariable efforts to escape detection that lead to the frequent instances of bites by this species. Cows are often bitten on the udder, and bulls in the same general region, and although it should be conceded that a puff adder pinioned underfoot or by the posterior part of the victim's body might possibly strike so high, it seems more plausible to believe the bitten animal had lain down on or beside the snake, and so incurred the usually fatal bite.

In tall grass or even in a well-grazed or trampled field it would be possible to walk onto a snake concealed in its habitual manner, even if it were known that the snake was somewhere within an area as small as four square yards. So far as my fairly extensive experience with rattlesnakes may serve, it seems safe to say that there are hardly any rattlesnakes that are more competent than the puff adders in hiding themselves.

The probable mortality for untreated bites is reliably stated at about 85 per cent for cattle, 100 per cent for dogs, and somewhere between these values for human beings. With serum therapy this high rate is apparently lowered to only a few losses in cattle, and many dogs and human beings recover from otherwise fatal bites.

Because of the high toxicity and large volume of the snake's venom, and because of the animal's habits and near-invisibility, it certainly constitutes one of the most serious threats to persons whose work leads them into the fields and weed-grown gardens, in fact anywhere but in deep bush or on open roads. Certainly every surveyor, hunter, prospector, or hiking farmer should wear leather protection as high as the knee, but it is obvious that this is neither employed nor likely to become popular. For all these men the usual clothing consists of shorts, socks or stockings, and low shoes, and except for the presence of viperine snakes, this would be the most practical garb in a humid climate with hot summers.

It is fortunate that puff adders are so cringingly shy, and that they apparently refrain from biting even when they are within a few inches of an inadvertently tramping bare foot or shoe. Why such dangerously venomous and comparatively robust animals should be so timid poses an interesting question as to their natural enemies, the predators that might search them out, or the part that man has played in killing all snakes whenever aware of their presence. There must be some reason for their exceptional timidity.

Although at one time semiaquatic and aquatic reptilian and amphibian fauna including the Nile crocodile were well represented in the rivers of Natal and especially Zululand, there is little reason to believe that even with protection any of these animals could have survived except in some of the largest bodies of water. In almost all rivers, sand and gravel from high up on the drainage slopes is now slowly filling in the old pools where the river fauna once lived.

Except for relatively short distances back from the mouths of the rivers, or in streams where drainage basins are almost wholly under the white man's control, sand has already made conditions impossible for most aquatic life. It is only a matter of time until sanding of rivers and streams will make life intolerable except in local areas where the violence of the floods is directed into channel scouring rather than this almost universal filling. The pool-dwelling life, piscine, amphibian, and reptilian, cannot survive for long because most of the old fauna was simply not adapted to periodic floods of mud followed by drought. Also the beautiful Cape otters will have to go, and also the aquatic birds, the little reed-swamp birds and reed-bed cormorants, ducks, anhingas, Peters' finfoot, and others. Only a very few of each may survive in areas of limited extent, and the once-populous fauna has even now dwindled to only a fraction of its original numbers. It was not the gun nor traps and snares that eventually brought disaster, but the increase of human numbers and its inevitable sequence—cultivation, erosion, death for the local animals.

TERMITES

The discussion of the aardvark and the monitor lizard drew our attention to termites. These fascinating insects deserve a few pages of their own.

During the first warm rains of spring and summer, life emerges from the earth where myriads of insects either have been resting as pupae or in hibernation; birds call from dawn to dusk, and happiness seems to fill the air.

Throughout the first few weeks of these warm rains and the accompanying steaming humidity, first one, then another species of termite begins to flutter up from the ground in nuptial flight. They rise almost vertically into the air, from their clay-and-saliva-cemented homes on delicate, transparent wings that make an audible rustle like that of soft silk. This is an aerial venture of romance, their first and last trip from a subterranean abode and the earth, which throughout life holds them in close embrace.

As they rise into the tepid, dark night, or the balmy light of day, each species according to its nature, gentle breezes carry them off into space, into a world they have never known, away from the protection of their fellows and the sheltering nest where they have been attended and waited upon all their lives. This drastic adoption of a never-experienced way of life is nature's way of assuring matings with other stock, a recombination of varying heredity lines and a distribution of colonies so far apart that only the most cataclysmic disaster could destroy the species as a whole. But this method is unbelievably expensive in wasted lives.

All night long prowling animals gather to the feast. On the ground, squat toads hop about wherever lights attract, or eddies of wind drift the termites in numbers, and there they cram themselves until their squat bodies are bloated and heavy. They gorge until their stomachs and finally their mouths are full, and one can see them reaching up with stubby forelegs to tuck in still more termites and to rub away the protruding wings that seem to tickle their lips.

In the air, bats dart back and forth, squeaking in high, almost inaudible notes, while nightjars flutter back and forth on silent wings, too busy to utter a sound. There must be many other forms of life busily engaged in gathering the harvest, for the bush seems full of rustlings and faint squeakings of rats, mice, and shrews as invisible animals scurry about in search of the termites that have already fallen to the ground.

Not only do the more lowly forms of life prey on the helpless termites, but man also takes his share. As soon as the *izinhlwa* start to fly, the message-sending falsetto voices of natives spread the good news from kraal to kraal, and these voices can be heard in all directions. Soon the black hills are dotted with blossoms of orange light, where fires have been kindled to attract the flying termites. As the insects swarm to the light, children and adults gather about the fires, catch the termites, and throw them into pans of water. This wets the wings and either renders the termites helpless or better still induces them to shed them. Hundreds are caught to be fried in their own abundant fat and eaten in handfuls. A few are devoured raw as soon as caught, the natives first crunching the head to avoid having their tongues bitten by recalcitrant insects. When fried, these night-flying termites are about the size of large pine nuts, and have somewhat the same flavor. When raw, they are distinctly different from any other meat, but are highly palatable to those able to forget that they are insects.

On the day following the flight of termites one finds pairs of the now wingless insects, scurrying over the ground, tandem, the one behind touching or almost touching the one ahead. During this time the diurnal birds and animals take a heavy toll of those that have survived the nocturnal carnage. Monkeys, herons, weaverbirds, in fact, almost every animal of the bush and field seems bent on getting its share of the rich food. As the day wears on, fewer of the frantically hurrying termites are seen, and here and there minute, tell-tale burrows show where a pair have taken refuge in the ground, their sanctuary from aerial and surface hazards. They will never see daylight again.

When all have gone, and more warm rain has fallen, all over the plowed fields bright red flowers come into bloom. The natives, seeing them, say: "Ah, there is a spot where the lovers buried themselves. When they dig into the ground in pairs, they do so to die, but they are only dead a short time, for there, you see, is the red flower into which they turn after death."

A smaller and more common species of termite begins its nuptial migration in the *daytime*. From observations on this form, it is easy to visualize more clearly what must take place at night when the nocturnal varieties emerge.

The individuals of this species rise from the termitary in flutter-ing lines that rapidly break up a few feet above the ground where even the gentlest currents of air carry them away into the distance.

The destruction of breeding stock in these diurnal termites is almost incredible in its extent. As one watches the nuptial flight, one sees, starting on the ground, the gathering of ground-living enemies. As the termites struggle to rise into the air, toads and lizards assemble and feed as voraciously on them as they do on the night-flying kind. Sometimes the assemblage of terrestrial animals is large enough to prevent more than a scattering of ter-mites from escaping into the air in flight. Many species of birds and even monkeys gather on the ground and take their share, feeding not only on the sexual forms, but on the workers as well.

When the termites do succeed in rising above the ground surface, they meet an even more relentless pursuit. At times the destruction is so complete that as one watches, hoping to observe the success of at least a few individuals, one is impressed by the fact that for minutes at a time not a single one escapes from the immediate vicinity of the nest. This is probably one of the most hazardous stages of nuptial flight. Near the ground predaceous insects as well as other animals compete for the rich food supply.

Time and again, as a termite rises a foot or so into the air, it is surprising to see it falter inexplicably and then fall back to the ground. On examination one finds a fly about the size of an or-dinary housefly busily sucking out the body fluids of its hapless victim. These flies, a dull brown in color, apparently appear from nowhere, and move with such speed as almost to escape detection. They pounce on flying termites and carry them to the ground, where they complete their meal at leisure. Robberflies, very similar to our North American species, also add to the destruction. Some-times one sees both termite and attacking fly, a double dividend, disappear into the mouth of some fortunate bird. Even the black, ordinarily nocturnal crickets emerge and feed voraciously on the helpless termites.

Above these lower levels, the most spectacular attack is carried out by birds of several kinds including the most unlikely species to find "hawking" for prey on the wing or stooping to consider such a comparatively small insect as a possible source of food.

As the termites begin to escape the earthly hazards and rise into the air the birds flutter out from nearby trees and dart about, catching their prey at random, but if the termite flight continues and is a heavy one, this haphazard method soon changes to a more systematic hunt. The birds become so numerous at times that they interfere with each other. This condition does not ordinarily occur until the swifts and swallows arrive on the scene and start pursuit of their common prey.

As these rapid flyers join the ranks of the bulbuls and other ordinarily frugivorous birds, the scene changes. The swifts sweep down at tremendous speed, catching their prey with such éclat that the click of the insects hitting against their bills is often plainly audible. Such speed, because of the large numbers of the birds in the air at once, creates a dangerous situation, for it might easily end in a mid-air collision with disastrous results. Probably with no reason but the terrain of the locality and the need for caution in flying, the darting birds circle around and around the spot where the termites are fluttering skyward so that in a short time most of the birds, especially the fast-flying swifts, are moving in a circle, one sector of which passes over the spot where the insects are most numerous. Often swallows (as many as five or six species) join the feast, but they are plodding flyers as compared with the swifts, and often they persist in departing from the flight pattern. Then they resemble little old ladies in fast traffic, bewildered, ducking and flinching as the business-like speeding swifts tear past them. The sight of this merry-go-round of birds is unforgettable.

Sometimes a hundred of the birds gather together, and the rush of air through their wing feathers makes a continuous hissing sound. Around and around they go, at tremendous speed, once in a while cutting across the circle or darting away from it in pursuit of some erratic termite that has not followed the regular line of ascent. Ordinarily the circle is fairly well maintained by the swifts, somewhat less so by the swallows, and only to a small extent by the less accomplished flyers. The clumsier birds anxiously flutter about over and under or around the circle, catching as many termites as possible after the more dexterous flyers have taken their share. Now and then an awkward hornbill flops clumsily through the mob of birds, catching some of the little insects in its immense and cumbersome-looking bill, or a hammerhead "stork" barges through the circling flyers, temporarily disrupting the flight pattern.

Around the outskirts of the fluttering, squabbling gang of birds, the drongos dart and swoop, oblivious of all the rest, relying on their speed and strength even in the midst of such a milling mob.

Although there is some chattering during these flights, most birds are remarkably quiet. Probably all of them are too intent on feeding to make use of their voices. There is even very little fighting, for the whole effort seems devoted to making the most of the brief but fortunate event. As the flight of termites weakens, the circling birds gradually drop out, one by one, until only a few of the more phlegmatic, or the still-hungry late arrivals, hopefully await a stray insect.

Some of these diurnal termite flights may last for hours. Depending on weather conditions and the numbers rising simultaneously from adjacent termitaria, the pattern of the avian termite-gorging will vary. If there is a strong breeze blowing, the helplessly fluttering termites rise into the swirling currents and are swept into the air, where they are soon widely dispersed.

At these times the swifts circle randomly through the air, high above those points where the hill configuration funnels air and termites into a rising vortex in which the swifts shrill their high-pitched, wild, ecstatic cries as they assemble to feed at these points of concentration.

Elsewhere other species of birds rise in steep flight to attack individual termites, their flight from the top of a bush or tree being an inverted acute V, at the top of which may be seen the fluttering detached wings of their prey.

Barbets, starlings, weaverbirds, pin-tailed whidahs, even the tiny little seed-eating frets, in fact almost every species of bird seems bent on glutting to repletion on this ephemeral bounty from the subterranean chambers of the earth.

Sometimes windy eddies drift scores of these aerial plankton into quiet backwashes of the air, and here they circle about hunting their mates and dropping to the ground. The speed with which a male doffs his shimmering gauzy wings when he finds his goal is startling. The shedding act is often so sudden that two sets, eight wings, lie dotted about on the surface of the ground in clusters.

Termites may rise at times when heavy rain would seem to threaten even their short period of flight, yet they are often seen fluttering upward through the torrential downpour, oblivious to all but the urge to flutter to their destiny.

On warm, quiet days following rains, when the ground is still soft and amenable to the feeble burrowing powers of these sexual individuals, a flight may start soon after dawn and last until midmorning. When such a day of emergence is under way, the air across a valley may be spangled with uncountable thousands of glittering wings as far as the eye can see. Against the shadowed sides of the dark bush-clad hills, each fluttering termite seems a sparkle of light against the gloom.

With massive flights such as these, when millions of sexual forms appear simultaneously, the enemies are simply gorged to such a degree of repletion that the aerial flight stages attain, for a time, almost complete safety. The termites have won survival by flooding the air, by overwhelming the enemy by sheer numbers, by incredible slaughter of the luckless with salvation for the lucky, and it is chiefly after such overwhelming flights that enough escape to ensure the founding of new colonies. Nevertheless, even underground, enemies in the form of bacteria and fungi, probably viruses also, take their toll, and drought and flood operate remorselessly from time to time.

The profligacy of nature with the life and feelings of the individual in order to assure perpetuation of the species, is beyond comprehension until it has been seen in a flight of termites, and yet this extravagant expenditure of individuals and emphasis on reproduction in order to assure survival of the species is not alone an attribute of the termites, but is true of all other life: bacteria, fungi, algae, the world of plants, and of animals, whether fish or amphibian, bird or reptile, and of the mammals including man himself. It is the tried and true way of avoiding extinction in years of drought or flood, season of plague and epizootics that strike far and wide; always, man and his fellow creatures have had a capacity to outbreed disaster. Those that became extinct, those that we know only as fossils are the ones that failed to outbreed a high death rate. If the death rate alone increased, and the birth rate remained static, extinction would follow.

Termite sexual flights are merely dramatic episodes demonstrating the seeming callousness of nature toward the individual. Only because of the obviousness of what is taking place are we appalled at the thought. For other animals the death toll is less conspicuous because it is spread over time. The monkeys skittering and skidding over the ground (out in the open for once) in pursuit of the termite delicacy are no more conscious of the ultimate significance of this sacrificial outpouring than are the native boys gathering to the same feast, or the birds that soar swiftly overhead as they mete out death to their victims while eluding their own inevitable fate.

After watching these flights one wonders that a termite species subjected to such hazards can maintain itself over the long ages that it has already survived. It seems that a needlessly wasteful method of ensuring distribution and cross-mating has been developed. Yet, when one travels over the country and notes the numerous termitaria it is obvious that by some means enough of the breeding forms do escape the dangers, and that after running this concentrated gauntlet, they succeed in evading the dangers to which they must be subject after shedding their wings.

On one occasion the termites were indulging in their nuptial flight in such numbers as I have never seen before. The air was cloudy with them in some places, reminiscent of locust migrations in the olden days.

There were no circling flights of swifts and other birds, the termite numbers being simply too great, and the rising swarms too widespread to favor that dramatic merry-go-round of gluttony; instead, the swifts, present in hundreds, grouped together for sociability or in rising warm air currents, where the insects rose to their waiting fate. There was the same rushing flight, the subdued murmur of black, fluttering pinions, the click and the silent fall of four wings, mute testimony to the speed and accuracy of this pinpoint capture at high speed.

In flight the termites resemble nothing so much as myriads of helicopters borne on glistening, gauzy wings, and like helicopters, they hover gently to a landing, in mating: the female first and then the males. Any miscalculation of distance or a sudden wafting breeze sends them close to the ground, where accidental contact sometimes tosses them onto their backs and, if the ground is wet, they may be held in place with wings stuck and legs waving desperately for an object to grasp. Even in this position, the wings are not always shed immediately, but it is interesting to watch the effect of a grounded female upon the hovering helicopter-like males as they drift back and forth a few inches or feet above the earth. Some hedonic odor must emanate from the female, for down-wind the fluttering males cross her invisible airborne track, swing back in line or overshoot, and by trial and error soon a small assemblage of males nears the passively waiting queen-to-be.

These flights are the consummation of a long period of preparation from egg to adult. This cinderella performance comes but once in the life of each male or female, he to die shortly, she to become a grossly bloated egg-laying machine. Yet there is no great temptation to remain in flight, and in this brief period, the female, after drifting through the warm, fragrant air for a few minutes, slowly settles to the ground, where she folds her diaphanous wings and waits for her destiny.

The female termite must exude her odor chiefly as she comes to rest, for in the air there is no sign of congregating males. As she sits quietly trailing her enticements, there is no conspicuous visual stimulus, but as she feels the male's first gentle touch and approaching ravishment and her farewell to sterile virginity, she lifts her "shoulders" with a seeming shrug and instantly divests herself of her gauzy aerial habiliments. He also, on the thrill of approaching nuptial contact, divests himself with equal eagerness, and together, tandem, with the male maintaining contact close behind, they seek the warm moist earth where in privacy their union takes place.

Their "day in the sun" extends only from their departure from an earthy nursery and long-time home to a moment's ecstasy in flight on the fairy beauty of their silken wings, then back to earth again to die or to undertake the task of starting still another colony. This is their life.

Within the old colony, a colony of years' or possibly centuries' duration, there is an accumulation of parasites and inquilines. A new colony minus this burden is started by the lucky pair with the advantage of new vigor from the sexual forms and new experiments in colonization. There may be only a million or a billion chances to one that any one female will be fertilized and survive to establish another subterranean cavern-home of termites, but, through most of their history this has been their custom. When at last abandoned, some kinds of termitary become a shelter for snakes and lizards of many kinds, including the black mamba, or sanctuary for swarms of honey bees.

At times there are extraordinarily dense flights of small termites, which take place during the rains following a thunderstorm. At this time they crawl under doors, through cracks between upper and lower windows, and swarm over and into furniture, lights or even the typewriter. Outdoors when a flashlight is trained on them, they flutter helplessly up the beam and then crawl up sleeves and pants' legs and down one's neck. When car headlights are turned on and they flutter in front of the lights by hundreds and thousands, the clouds of little helicopters wax and wane as breezes fall or grow stronger. After thirty minutes, the ground becomes littered by a solid mass of wings and crawling insects.

The question as to what factors precipitate the flight of sexual termites is still a puzzle. One Thanksgiving evening for instance, under moderately dry but warm conditions, there was a fairly heavy flight even though many days of dry weather had preceded it. Then, on November 29 after two days of rain, at first cold and then warming, termites again emerged in large numbers.

The probable inducements or stimuli that precipitate the emerging flight of these insects must be closely integrated with the underground cycle of egg to maturity; egg laying and nurture are meticulously adjusted to favor flight at the opportune moment, but the trigger mechanism for the actual moment of the springing loose of the waiting hordes of sexuals may be based upon both the humidity of the air and its temperature combined.

Since there are both nocturnal and diurnal forms, the intensity of the light may serve to stimulate the day-flying forms by a transient reversal of normal tropisms, and lack of light may determine for the permanently photophobic species those that emerge at night.

Many winged individuals appear at the mouth of the termite burrow, protrude their heads, then seemingly test the air and the above-ground environment, turn, and retreat into the termitarium. This is particularly notable toward the end of a flight. Occasionally one sees the workers grasping these sexuals by the wings and tugging them back into the burrow.

For insects that have spent their lives underground and are normally photophobic and highly dependent on the extremely uniform climate consisting of humidity, darkness, and temperatures encountered deep below the surface, the temerity of these diurnal flights, and nocturnal emergings as well, this bursting out into a totally strange and different environment is most extraordinary.

Adult sexual forms must require a remarkable adjustment, a brief transformation of normal termite physiology and psychology, or they must detect and respond to rare and exceptionally favorable external conditions that closely stimulate the environment within their home shelter.

It is probable that in their evolution the sexual forms represent a combination of temporary change in neural and physiological functioning, and also that they thus automatically select the most favorable time for emergence. Even the physiological change that keeps them fluttering is most ephemeral, as it lasts for the colony only a few hours at the most.

Although the workers and soldiers of most termites are slow moving, they are exceedingly interesting because of the division of labor and the specialization of members of a colony for performing separate duties. Specialization has been carried so far that not only in function but also in body form, the castes are often easily recognizable at a glance.

In one genus, *Nasutitermes* (*izinkulungwane*), the head of the soldiers has become elongated into a slender, forward-projecting snout resembling an ancient helmet with long, sharply pointed beak projecting straight forward. It looks like a formidable weapon of defense, at times possibly even of offense. Under the surface of this beak, which is heavily armored with hard chitin (insect "bone"), there are glands that secrete a viscid, sticky fluid, and ducts that carry the substance out near the tip of the snout.

It has been said that the substance is merely an entangling fluid, but my own observation as a small boy suggested that it may be poisonous, at least to each other. The liquid is not harmful to larger animals, and even though the termite warriors may swarm out over human hands, they fail to inflict perceptible wounds. Even on the tender skin between the fingers, they seem unable to do any damage.

To watch the soldiers in battle it is necessary to repeat a youthful game, learned from a small herder boy, and start the conflict by artificial means. This is easily done by taking a large fragment of one termite nest, after the soldiers have swarmed forward to its defense, and carrying it to another termitarium some distance away, where the soldiers are dusted into a break. Sometimes it takes only a few minutes to start what looks like a fierce war, and the onlooker becomes absorbed in watching combat on a Lilliputian scale. However, there are many inexplicable occasions when there seems to be little, if any, warfare.

At the first sign of danger the workers retire to the deepest recesses of their "castle," and the soldiers swarm to the breach. The battle presents a picture which combines the legends of centaurs with the age of chivalry, but these centaurs are six-legged, and the suggestion of knights of chivalry comes only from the armored heads. For all of their six legs and the similarity to either centaurs or knights, the combatants move at a slow pace. When battling, they joust with their heads, and when they "set their lance at rest," ready for the charge, they can only settle those immense heads more firmly on their "necks," and slowly amble forward to lunge at their opponents. Agility in termites is notable for its absence.

When grass fires have swept the pasture lands bare, one of the features of the Natal landscape is the abundance of termite nests of *Nasutitermes* which rise to heights of twelve to eighteen inches, and dot the field with their rounded masses. The similarity between these nests and the old-style Zulu huts is so great that one might almost think that the Zulus had taken them as a prototype for their homes and erected their domed huts among the small miniatures made by the termites. It is this species of termite that provides the incubator for the eggs of the Nile monitor.

These are the termites that build fungus gardens underlying the dome-shaped, protruded part of their nests. Grass seems to provide most if not all of their material for the garden, and they must be very effective agents for accelerating the decomposition of grass into its various components, beyond the action of the elements. Owing to these and many other types of termites, some of which rapidly enclose and destroy twigs, stumps, and logs lying on the ground surface, soil building and cycling of the elements must be speeded to a rate that would not occur under the more familiar processes of bacterial decomposition.

It is interesting that the singing of the little "toad" *Breviceps* ends simultaneously with the end of the flights of termites. Other species of amphibians become audible even later, but these are marsh singers, and with habits quite different from those of the clumsy little ground-burrowing *Breviceps*.

It seems quite possible that over the long passage of time this little terrestrial burrowing toad has integrated its major activities and life cycle with the abundance of food provided by the enormous flights of the termites, and that it may even have become dependent on them during the energy-sapping season of reproduction. Their mutual requirements of high humidity and moderately warm temperature, and their largely subterranean existence throughout most of the year would seem to favor the relationship of toad and termite.

Additional studies concerning the adaptation of various species of African animals to the life cycle of others should be rewarding. Even man's welfare may be related for good or bad to soil conditioning by these abundant insects and the rate at which they reduce organic matter and mix it with the soil. Certainly their ecological significance to the Nile monitor, and to bees, snakes, and rodents constitutes a challenge to research.

7

NIGHT AND SUPERSTITION

As far back into the history of man as folk tales and written records can take us, superstitions have served as a substitute for knowledge. In the long process of man's gradual and stumbling progress from benightedness toward intellectual emancipation, that is, his escape from metaphysical explanations of his world, mere belief has never been successfully substituted for knowledge. Even Western civilization all too often still relies on faith in order to fill the vacuum resulting from our ignorance. The African, almost completely captive to his superstitions, has a still longer intellectual journey ahead of him.

Contact with the representatives of Western civilization and the driving forces of world change is prematurely ripping away the protective shell of the African's ancient beliefs and practices. His position is analogous to that of a moth's when its chrysalis is forced open prematurely; damage that leaves a scar is the inevitable consequence. This alone might not be so serious were it not for the fact that while casting off his old shells of bondage, he is adopting many of the shackles of superstition that continue to keep the West in darkness.

It would be far easier to keep one's faith in Africa's future if there were evidence that both the African and the white had greater trust in knowledge than in superstition. In addition, our confidence in South Africa's future would be fortified if we could foresee a century of time for the evolution of acculturation, but both the speed at which populations are growing, and the only alternative, the foreseeable ways in which this might be brought about, forecast revolution rather than evolution.

In general I have found Americans more willing to laugh at the African superstitions than to take an interest in them and understand them for what they are. This is particularly amusing as we witness the individual faith of many Americans in the rabbit's foot, a lucky medallion, or other talismans against night and danger, or even in the rapidly developing blind faith in "science," of which so many of us know nothing. Yet we are willing to substitute faith in science for hard thinking and the process of learning.

The native's superstition was born of his fears, especially those aroused by dangers, real and imagined, that lurk in the darkness of the African night. Before the moon has appeared, the bush seems less alive than it does with even a faint lunar glow, but in the darkness there is a feeling of suspense. At these times the enveloping globe of night that circles every campfire and even the lamp-lit houses makes the surroundings seem hostile. For the natives in their lampless kraals, the night presses in from all directions, and for them there is no alleviation of fear of what lives in the outer darkness. The little rustlings in the leaves, the sharp squeak of a rat, or the flutter of wings endow the night with life and mystery.

When the sun has set and the hills turn to irregular silhouettes against the darkening sky, Africa becomes more than ever a land of mystery and beauty. During the nights, the dark is alive with strange sounds. Of course, the voices of the larger mammalian species are gone forever from Natal; the sawing grunts of leopards, the rumbling growl of lions, and the usually melodious calls of hyenas have long since vanished from these regions, and now it is only in the sanctuaries and on a few large farms that they linger on by rare acts of tolerance in man. Zebras and the larger antelopes have also gone.

To go for night walks in the vicinity of the missions was my supreme pleasure during my return visit in 1953. Just at dusk the nightjars would whistle their whippoorwill-like notes, and on every dim path, their shadowy, silent forms would rise and lilt along before the stroller.

The most common nightjar is the *usavolo,* a bird that the native medicine men or herbalists use in preparing a soporific, especially designed for children that are fretful at night. The Zulus, not differentiating among the species, apply the name of *usavolo* to all members of the nightjar family.

It is difficult to understand why a bird that is wakeful at night should be selected as a sedative. It carries on most of its noise-making after sunset and before sunrise when children should be asleep, singing vociferously only during the hours of faint light. On moonlit nights, however, *usavolos* may be heard calling all night long, but even then their chorus is most in evidence during the borderland hours between night and day.

When the balmy, fragrant winds of night blow gently over the face of a restless sleeper, insomnia is not an ordeal as it is in cities. In the darkness of the lonely bush most sounds are musical, and there are no squealing tires and brakes, crash of traffic, raucous honking horns. Unless one is utterly worn out, it is a pleasure to lie awake.

Even the little Woodford owl, *mabengwana,* which the Zulus fear as an emissary of a dangerous witch, is soft-voiced and musical. As the call, "woz', woz', mabengwana" (come, come, mabengwana), echoes through the dark bush, the notes blend superbly with the other night sounds. Far away a Zulu, singing his simple, barbaric cadences, may add another natural note to the general pattern of sound, and will seem no more an intruder in the wilderness than the cricket that is singing so energetically under the grass nearby. The crickets of any country have received their well-deserved praise from poets and naturalists, but the "voice" of one of the African crickets ranks as an insect song supreme. Although simple in cadence, no sound in nature seems more cheerful and, in its surroundings, so serene and dauntless. In the vast night that makes the world seem so overwhelmingly fraught with danger and that dwarfs each individual life under its folds, this simple song rises lustily, steadily, unafraid, although protected by only such humble shelters as a rotting log or a few blades of grass.

Although night may be the time for other senses than vision, there are still some evenings when nature produces a lavish display of enchantment, a fairy scene in which fireflies play the roles in an extravaganza of darting, scintillating pyrotechnics that only the dullest eyes could fail to enjoy.

In early summer, late November or early December, these insects emerge in hundreds and thousands, and in every low-lying piece of ground or in the remaining vestiges of bush, the blackness of night, or early evening before moonrise, becomes a tapestry of moving futuristic patterns, threads of fire against black velvet.

When a warm wind is blowing they appear in their greatest beauty, for then they drift and accumulate in protected hollows or along the lee of taller vegetation. In their thousands, darting erratically in crisscross patterns of fire, they suggest more the sparklers of our Fourth of July than approaching Christmas.

There seems to be little doubt that one reason for the brilliant luminescence of these night-flying insects is sex or species recognition, and that the flights, like those of the termites, end in mating. In the darkness of the night it is difficult to follow individuals and to study their behavior in flight or on the ground, where some of these major displays end in dramatic abruptness. It may be true that with their strong odor there may be mixed some fragrance of hedonic glands as well as flashing lights to aid in the fulfillment of mating, but in many species the lights may play only a minor romantic role or none at all.

If these insects that kindle their lights chiefly, or only, in flight, rely on odor for the consummation of mating, is it possible that some other, as yet unreported factor, may play a part in these conspicuous displays? Certainly without their lights they would be as invisible to predators as the multitudes of other insects that also drift to sheltered places in their flight. The predators that feed in darkness are the bats, those nearly blind, sound-dependent, sonar creatures, speedy of flight and hunting prey while on the wing. Others feeding at dusk or seeing their food against the sky glow are the wide-gaped nightjars.

These hunters on the wing must take their food while in flight at high speed, and while doing so presumably they must pick the good or avoid the bad. To our taste, or sense of smell, these self-advertising fireflies are thoroughly obnoxious.

When the night is swarming with insects of all kinds, the hunting by birds and bats would call for some instantaneous and positive means of identifying the good from the bad. Therefore, do these flashing lights of fireflies function not only as mating guides but also as clear, unequivocal messages to the bats and birds that here are nauseous inedible insects?

Where it is possible to watch the fireflies against the background of the night sky, it is immediately apparent that various species of bats are engaged in their erratic zigzag food-hunting flights among and between these sparks of light, and that the nightjars are also feeding; both animals supposedly catch the most delectable of the insects and avoid distasteful species, just as their diurnal counterparts have been proved to do.

In the daytime, brilliant coloration gives away those species that are distasteful, poisonous, or provided with potent stinging mechanisms. The vivid and conspicuous color, and the patterns associated with such colors, have become known as warning or advertising coloration. (The device is so effective that many species of insects mimic such forms and coloration for their protection.) For night-flying forms that are preyed upon in the dark when color cannot show clearly and give this protection, could anything be substituted more effectively than flashes of light?

This problem would be particularly acute for insects that are preyed upon by enemies that catch them while they are in flight, merely silhouetted against the faint light of a night sky, or still more difficult for the bats whose rapid zigzag flight and poor eyesight would provide them with little ability to discriminate between species. For the bats, some peculiar sonar reflection or sound might serve as an effective warning, so that, having experienced the sound and a disagreeable consequence, they might associate the two and thereafter avoid a repetition of a painful or distasteful experience. The nightjars might also learn a similar lesson by distinguishing between silhouettes. After having watched these animals while they were feeding where fireflies and other insects, especially the nearly universal food, termites, were numerous, I could not escape the conclusion that the fire-flashing abilities would provide by far the most effective warning at nighttime.

Possibly this is the basic value that has encouraged the evolution of such a spectacular structure as the light-emitting organ. An experimental testing of this possibility could be carried out with little difficulty, and would serve to add to our store of knowledge of both prey and predator.

At night the value of sight diminishes for mankind, and the senses of hearing, touch, and smell are keener. Ordinarily we rely on sight more than on any of the perceptive stimuli, and our psychological behavior is usually coördinated to a large extent with this sense. At night our reactions change, because of the suppression or absence of stimuli coming from the most accustomed source, and then we who are accustomed to light at all times, day or night, and seldom depend on any sense but eyesight, find that each little stimulus that would ordinarily pass unnoticed in the glare of daylight sinks into the mind and reaches the threshold of consciousness. When a dove awakens at night and gives his melodious notes, the sound does not seem merely the song of a bird, but to the sensitivity of half-sleep, and uncontrolled imagination, it becomes infinitely more. It suggests all the dangers of life in the bush, and the vast numbers of animals at rest, or the other vast host, the animals preying on those at rest. Similarly, the barely heard suggestion of sound from a fretful child in a distant hut brings vividly to mind a mental image of all of Africa lying asleep, the sleep of millions of huddling black forms, the life they have led, and the tremendous mass of unconsciousness flooding over the continent with the onward sweep of the shadow of night.

All forms of life that remain invisible and unheard during the daylight hours begin to move about and give their various calls or songs as soon as night falls. It is only then that one fully realizes the immense number of animals that carry on a furtive existence, unseen and almost unknown by man, and yet close to him at all times, day and night.

When the first rains have soaked into the ground, and the mist hangs low in the valleys, the nights become more vocal than ever. Thousands of little frogs, almost exactly like our tree toads, inflate their throats and shrill their loud, cricket-like calls. At times, when I collected specimens from the marshy areas where they are found in greatest numbers, the chorus was almost deafening.

The effect of the continued shrill uproar is exasperating at close range. It "feels" as though it actually emanated from within one's head, and eventually there seems nothing but this high, shrill sound, pulsating with the regularity of a slow heartbeat. Much of the effect must come from some mechanical reaction to the sound waves, rather than from the noise itself, and there may be some connection between the hearing mechanism and the noise's periodicity, characteristics, and phase effects from the vibrations. But whatever the cause, the sound creates a strong illusion that there must be thousands of little amphibians close around, whereas careful examination seldom reveals more than half a dozen at any one location.

From a greater distance the chorus is agreeable, and especially so when some of the larger frogs add their deeper, calmer notes to the harmony. Some frog calls suggest those of American amphibians, as do the makers of the songs themselves, for they are members of the same genus as our common meadow frog. But in the 'twenties there was one mystery voice, which I usually heard somewhat later in the summer. The pitch of this call note seemed deeper than that of our American bullfrog, and since size and pitch are correlated to a degree, the bass voice that boomed out only a few times in the course of an evening, suggested to my imagination some very large species as yet unreported from the area. The single resonant booms came too infrequently to allow my tracing them to their source and so serve to identify the animal, but the natives maintained that the maker was a giant frog, and from their descriptions it is entirely possible that in this instance their natural history was correct.

Since my efforts to identify the source of this voice failed, and since in later years all marshy habitats had become filled with sand and gravel, the owner of this voice remains still another mystery of the night, and may never be heard again in that area.

During the early hours of the night and just before dawn, the giant fruit bats gather in flocks, especially at the season when the *umtombi* fig tree is in fruit, or when the loquats are ripe. Sometimes, usually at full moon, they seem to be present all night, and at these times and occasionally also on dark nights, one bat may begin to call. For hours a light sleeper may awake and listen unwillingly to the regular rhythmic beat of the eerie voice coming from the darkness of the lonely bush. Every few seconds the insistent note, neither loud nor faint, and metronomic in its regularity, beats on and on until the sound seems to merge with one's own being, and one seems to be dreaming of it through the queer fancies that come with malaria or other febrile upsets.

The unknown sounds of night add fascination to its beauty and intrigue the naturalist's curiosity. There is one notable sound that is so rare that it adds much mystery, and one puzzles as to how the riddle of its maker can be solved. On hot, still nights when the summer air is heavy with the perfume of nocturnal blossoms or with the portent of rain, from high in the air comes a husky whistle. There are fairly regular intervals between the notes, and apparently the animal's flight is direct rather than wandering, for the sound is ordinarily heard in the distance, then comes closer and passes overhead to fade again in the opposite direction. On one occasion in the steamy weather following a rain I heard the whistle approaching rather slowly, and there was time to get out into the open and search for the object. The sounds came nearer and nearer, and then for a minute they came from above as though the maker was circling overhead. I could not see anything against the darkness of a cloudy sky and the sounds became fainter and faded in the distance.

This particular sound is only rarely heard and then only when the maker is in flight. The Zulus claim that the sound is made by a grasshopper-like insect, and similar information was given to me by two white men. Yet, it is difficult to understand how the animal making the sound could ever be identified.

There is another unusual voice in the bush, sometimes heard in the daytime but usually at dusk and only during hot weather. This is a hollow, moaning cry that rises from the center of some patch of dense vegetation and therefore, although it may be repeated time after time and sometimes for minutes on end, it seems never possible to come within sight of the maker. The Zulus claim that it is the moaning call of the *imbulu* lizard, a monitor, suffering the agonizing pains of a toothache and with no hope of dental care, but imaginative as is this interpretation, and although the sound is suggestive of such agony, the monitors are incapable of moaning noises so far as is known, and a more probable explanation is that the call emanates from a migrant European bird, the corn crake, a member of the rail family. Whatever the source of the noise, the shielding cover is always so dense that the little animal, bird or reptile, can easily escape discovery, and while being sought has ample time to stop its moaning and slink out of sight.

The African night affords pleasures other than the sounds of nature. During the winter the crisp, cool air is permeated by a vast kaleidoscope of smells emanating from countless sources in the country, from the many aromatic herbs of the bush, or simply from the sweetness of dry grasses and a clean world. But it is during spring and summer when the humid air aids the sense of smell and new life is bursting into being, that the scents of night reach their intoxicating peak.

After the long dry months of winter the first rains bring that most nearly perfect odor, that of wet earth, as clean and sweet an odor as that of any flower. To lie awake at night and listen to the rush of rain pouring down on the hard, dry earth, and to sniff the damp, cool air eddying past, is a delight that only those acquainted with long periods of drought can fully appreciate.

As spring rushes on toward summer, the flowers begin to bloom, and it is at this time of the year that the bush becomes a continuous olfactory banquet. The so-called Cape jasmine, with the fragrance of a gardenia, blooms along all edges of the bush, and the white-faced, pink-backed blossoms seem to give their best at night.

On some of the cliffs that are not too steep to be partly wooded, a species of ground orchid blooms for weeks, and the air, especially at night, is heavy with its perfume. It blossoms almost continuously, each spike of flowers carrying the fresh white blossoms midway between the tips of greenish, undeveloped blooms and the older part with pale yellow flowers, those which have almost passed maturity. Lowest on the stem are the brown, dried ghosts of those that blossomed first.

Sometimes the orchids flower more heavily than usual, and at these times they form banks of bloom along the cliffs so great that armfuls can be gathered in a few minutes. Kept in the house they last well and perfume the air of a room with delightful scent.

Another orchid, rarer than the rest, and found clinging only to the trees, is small and modest and seldom seen, but its perfume reaches the wakeful at night. The orchid is a creamy white, in general shape much like a violet, and the edge of the flower is trimmed with a fringe more fragile than the finest lace. One blossom of this species delicately perfumes a room with a faint odor suggestive of vanilla.

But in Natal wild orchids are rare, and the more plebeian plants are those that count most heavily.

Being aware of the fragrances are some of the delights that have made of insomnia not something to be dreaded or avoided but simply a minor disability that pays greater dividends than the premiums paid. The dividends are more hours in which to absorb the sensual, aesthetic pleasures of life in the South African countryside.

ABATAKATI, IZINSWELABOYA, AND UMKOVU

The accumulated fears of centuries weigh down on the native, and it is no wonder that the dark holds overwhelming terrors for him. Even nonsuperstitious white men forced to spend a night in the bush, with no comforting fire and no protection from the oppressive darkness, have felt at times that back-chilling gooseflesh. In the dark, even today, there is actual danger from treading on a snake, the lone surviving dangerous animal in much of South Africa. It is not surprising, therefore, that night engenders fear among the superstitious natives; furthermore, these people have just emerged from the age of general mammalian dominance, from a time when predators were the kings of the wild and man was merely another kind of diurnal mammal with poor night vision.

It is immaterial whether the native is afraid of the dark or the things that it may cover. Probably the night of itself holds no fears for the ordinary mortal, but to an imaginative person darkness becomes something that shields limitless possibilities.

According to the Zulu superstitions, night belongs to the *abatakati* (male witches, wizards) and the *izinswelaboya*, professional murderers who kill for the sake of human parts with which to make "medicine." It is during the darkest nights that the wizards ride on their trained baboons; the word *imfene,* or baboon, to the Zulus conjures up the utmost in evil on these weird night rides. They believe, and the belief can stampede them wholesale from any place of employment, that in the distant hills where the bush is dense and black there still live wizards. Even two or three generations of education have failed to exorcise these fears. So, even now, most natives believe that if these evil ones are properly invoked by some enemy and are well paid, they will accept the task of bewitching anyone.

Although there are many simple ways of bewitching, the natives hold that a favorite method of carrying out these dire errands is to train a baboon to carry the deadly "medicine" (*muti*) to its destination. Even more exciting is the common belief that the wizards often prefer to pick out a big, powerful baboon which they mount astride, and then at a word of command, the baboon will fairly leap through space and arrive at its destination with supernatural speed. While straddling these baboons, the wizards must always ride facing backward, holding firmly to the animal's hair, as they are carried about on diabolical errands. There is not the slightest use in combating these *abatakati* attacks. The baboon will always warn of dangers ahead, and since the wizard invariably rides facing backward, anyone attempting to follow is certain of being seen. No one has ever ventured to obtain vengeance on rider or baboon.

The natives fear even the bare mention of this dread witchcraft, and it is usually difficult to get them to talk about it. This is especially true on the white man's farm, where the owner is either scornful or unduly curious concerning native customs or beliefs. The better-educated natives are more sensitive to ridicule on this subject than on almost any other topic, but when at last their confidence has been won, they sometimes pour out a marvelous collection of tales.

One of the most dreaded of all night dangers is that of the emissaries of the *abatakati*, the *umkovu*, having a human origin. They are possibly even more to be feared than the baboon.

It is said, and probably correctly, that one reason for hiding a relative's grave is the fear that the grave will be opened by the wizard, the body exhumed and converted into one of these dreaded *umkovu*.

The procedure for a wizard is to exhume the body of some unfortunate native, cut out his tongue, pour a variety of concoctions down the corpse's throat, and in other ways prepare for the consummation of the act. When all is ready, an iron, which has been heated to red-heat in a nearby fire, is pushed downward through the top of the victim's head. As the iron goes sizzling into the flesh, the corpse shrinks accordingly, and when about two feet high it is ready for the final act of coming to life again. This may happen spontaneously, or the wizard may have to blow dust into the nostrils of the corpse.

Whatever the ghastly resurrection procedure, as soon as the now dwarfed figure comes to life, the eyes begin to glow as though a flame were burning behind them, and the little body, forever devoid of volition, becomes the perpetual slave of the wizard.

By using an *umkovu*, the wizard simplifies his work to a great extent and is perfectly safe from detection. All he need do after obtaining one of these slaves is to send the little dwarf with the glowing eyes to the hut of some luckless native. As long as the door stays shut and the inmates do not answer to a knock, they are safe.

If some person inadvertently opens the door to answer the knock and sees this little man standing there, he either drops dead or becomes dumb or blind. If a person vocally answers the knock at the door, dumbness is sure to result.

It is probably due to a belief in this method of witchcraft more than to any other reason that Zulus cheerfully chatting inside a hut at night stop talking and become absolutely silent when someone knocks at the door. It is also the reason why they will not, or at best most reluctantly, answer the hail of a stranger at night. It further explains why they are so unwilling to go to a neighbor's hut after dark to answer an emergency call.

It is rare to see a native walk cross-country at night to call for help even in an emergency. The imagined dangers of the dark outweigh the real ones of snake bite, accidents, or, in the past, lions or leopards. An emissary who will brave the dark even for a friend is rare indeed and is a testimonial of devotion far beyond normal; there are such heroes but they are not often seen by a white man. In one such instance the man who needed help had been thrown from a horse a month or two previously and received some damage to his head. The injury, though not visible, caused a good deal of distress, especially in the form of violent headaches, so he had hired a native doctor who had then proceeded to *caza* him on the head, that is, performed a blood-letting from a vein or artery above the eyebrow in the hope that the flow of blood would "carry off the sickness." The operation had been performed botchily, the blood vessel being opened not by the old skilled method, a longitudinal incision, but by a clumsy transverse cut. After the treatment—performed of course, with handy pocket knife and without any antiseptic—the wound had only partially healed, but the man paid little attention to the small and hardly noticeable wound. At length, however, a sudden fit of coughing burst open the cut. Infection had weakened the local tissues around the wound, and only a slight strain was necessary to break it open. The blood squirted from the man's wound, and he was unable to staunch the flow. By the time he had prevailed on a friend to ignore superstition and go for help, he was weak and soon afterward fainted. The friend decided to call me.

The wounded man's wife, frantic with fear, attempted to stop the flow of blood. First, she wrapped her husband's head with the leg from an old, dirty pair of trousers, then she placed an even dirtier rag, and when at last she was absolutely certain that the blood would not penetrate these cloths, she brought out a brand new bath towel from her treasures, and, with the price tag still tied to one corner, carefully proceeded to cover up these dirty rags and make her husband comfortable.

If he had been conscious at the time, he would doubtless have approved of her care in keeping the nice, clean cloth unstained by blood.

Fortunately for herself and her husband, after putting the cloths around his head, she shifted them slightly and by accident happened to bend the projecting end of the blood vessel backward on itself, as one would double back a flexible rubber hose to stop its flow. This had doubtless saved her husband's life, for it effectively staunched the bleeding. Except for this piece of luck, any help would have arrived too late, because it was necessary for me to ride miles back along the path that the friend had so valiantly taken earlier in the night, and since I did not know where the kraal was, it was necessary to allow the messenger to proceed at his own gait, much slower than that of my horse.

I will never forget that early morning ride. Knowing that a life might depend on speed, I kept the horse at a near trot, continually forcing the tired guide almost into a run. From time to time we plunged down into a mist-filled valley only to climb laboriously upward to another ridge. There was enough starlight to turn the mist a bluish gray and at intervals, where some lone tree stood out along the top of a ridge, the branches, blurred to unreality by mist and darkness, flung their weird shapes into the air.

For an hour we moved steadily along. Now and then a cock crowed in the distance or a dog barked suddenly close to our heels, but the rest of the time there was only the clip-clop of the horse's hooves and the shuffle of the guide's naked feet on the bare ground of the narrow native footpath. The ride was eerie; all the world was unreal, and at times we seemed to be the only human beings in a universe of gently whirling river mist. No wonder that the ignorant natives are afraid of the dark!

When at last we neared the kraal of the sick man, the trail had disappeared into the dense, black bush. It plunged almost vertically down a bank far too steep for horse or even mule, and seemed to melt into a pit of solid black.

The hut was of the old-fashioned kind and was set on the ridge beyond this stretch of bush. It was necessary to kneel down to enter the low door that led into a room dim with smoke, lit only by a little fire burning on the floor in the center of the building. Beside the fire the wife was crying quietly, rocking back and forth. Three little children clung close to her and turned to look at the strange visitor, possibly the first white man they had ever seen. All that was visible in the murky dark were the whites of the round, big, black eyes of the frightened, tearful, and bewildered children. At the side of the hut lay the man so near death. As help arrived, he managed a faint *"sakubona n'kosana,"* the polite greeting meaning, "I have seen you," then relapsed into unconsciousness. From where the sick man lay, all the way to the door, a great, fan-shaped stain of black discolored the earthen floor. In the little depressions on the ground were pools of black that had been flowing blood.

An hour's work with drops of brandy and warm stones brought consciousness again, and at last, just as the sky grew light, he seemed to recover strength. I administered antiseptics and gave him clean bandages.

A month later when I passed near this same kraal, the man shouted across the valley that he was well. The treatment, but chiefly his own resistance, had brought about complete recovery.

Undoubtedly, the fear of dangers lurking under the cover of darkness is responsible for many deaths that might otherwise have been averted, and it is rare indeed to find natives who for any reason will travel across certain notorious streams or rivers or through certain well-known patches of bush after dark. Where no particularly dreaded haunt intervenes, where there are no potential hiding places, the natives will travel somewhat more freely.

The only actual danger today—snake bite—seems to deter them only slightly if at all, but they claim that there are still the *izinswelaboya*.

As a small boy, I discovered that a white sheet and a well-made jack o'lantern were all that were needed to drive the big normally fearless men-servants pell-mell into the house, where they locked themselves in for the night. It was not feigned fear either, and small boys found it great sport, but it must have been real agony for the men.

Nowadays, the natives are becoming more sophisticated, and masks or jack o'lanterns merely bring out the remark "fokisi," that is some trick of Guy Fawkes Day, an English holiday and time of celebration. Nevertheless, these same apparently sophisticated Zulus would run in terror if they should meet such an apparition where no white men were known to be present. They still believe in the *umkovu*, even though *fokisi* has possibly complicated matters in the towns.

Umditshwa, a big, muscular Zulu weighing nearly two hundred pounds, was almost fearless in most respects, but he was typically Zulu in his fear of the *umkovu*. After working for the mission fourteen years, off and on, and having been exposed to the white man's influence throughout this period, he still remained at heart afraid of the dangers of witchcraft.

This great fear was even more surprising because for many years he had worked with a native "doctor" and had for a long time contemplated becoming a medicine man. He was acquainted with all the potent remedies for ills of one kind or another, knew the love philters in especial favor, and carried a complete collection of amulets to ward off the various dangers of sorcery. As a small boy, whenever I needed pennies with which to buy bullets for an air rifle or rubber for a slingshot, I could sell a few bats to Umditshwa, who used them in some of his most efficacious remedies. Dried and pulverized bats seemed to rank high in the Zulu pharmacopoeia, but I was never able to learn their specific virtues.

Bats' wings were also used as coverings for some of the horns containing little bunches of herbs, but the modern "horns" (glass vials) used as amulets seem to be considered equally effective although stoppered with the more prosaic corks of civilization or even bits of newspaper. Passing of the good old days must grieve the conservative "doctors."

Of all the Zulus, Umditshwa was the most reliable source for information on herbs, their uses, and names, and the names of trees and grasses. He was also well informed on the names of mammals, birds, and reptiles, and on the life histories and habits of the various living things around him. But in some details, super-stition was mingled with fact.

At night, when the dew was on the grass and the moon was full, Umditshwa was in the habit of gathering some of the rarer herbs for his collection, and at these times seemed to ignore the real and fancied dangers of wandering about at night. As he explained it, certain herbs were of value only when collected at night and with the dew on them. The condition of the moon may have had some significance in herb hunting (these herbalists are also known as *inyanga,* the word for moon), but it was never specifically men-tioned, and, according to his information, the light of the moon was helpful only for visibility.

As a brewer of emetics, a sovereign remedy with the Zulus, Umditshwa was unusually successful. One of the herbs that he particularly favored as an emetic was obtained only at night, and only while the leaves were wet with dew. After gathering, the herbs were carefully dried and eventually, just before using, brewed into a strong infusion, about the color of strong, black tea. This particular emetic was pleasant to the taste and was character-ized by a strong, mintlike odor. It was considered mild but thor-ough, and on one occasion, when I was feeling ill as a small boy, and Umditshwa happened to be defending his remedy of *palaza,* or taking an emetic, I secretly persuaded him to let me try it. It was difficult to induce him to brew the medicine, probably because he feared harsh criticism if he were found to have been administering native remedies to a white boy, especially as my mother was noted for her ability with the white man's medicine.

He brought me about two quarts and said that it would be neces-sary to drink it all, and then retire to a warm, private spot and sit with back to the sun until the remedy could be felt moving toward its destiny. Knowing that the Zulus required a heavy dose of European medicine to stir them, a smaller quantity than a quart seemed sufficient. It was—emphatically.

When completely free from the last drop of the emetic, one does feel better, not necessarily due to the medicine, but a seasick person also feels better after a violent attack that has been carried through to the normal crisis.

Some remedies known to Umditshwa were exceedingly vile in color, odor, and taste. Some others were fairly pleasant in odor, but as to taste, Zulu medical philosophy asserts that effectiveness is proportional to the vileness of a medicine.

It would be surprising if some of these herbs might not produce drugs valuable to our own pharmacopoeia, but it is equally certain that some remedies are based solely on superstition. On one occasion, when I had been stung by a small scorpion, Umditshwa immediately ground the beast to a paste on the blade of his hoe, then poulticed the site of the sting with this paste. Obviously, if there had been any poison left in the scorpion's poison gland, that also would have been added, somewhat diluted to be sure, to the poison already in the puncture. The pain soon left the wound, but I could not persuade Umditshwa that it would have disappeared even without his treatment.

Most of the legends that I learned directly from the natives came from Umditshwa, a real friend, and it is to him that many thanks are due for my introduction to the knowledge of native stories and medicine preparation. He also helped me to gain some slight understanding of some of the native fears and customs, thus enabling me to make friends with other men. Probably the most impressive story told by him resulted from my gentle ridicule of his fears of an *umkovu*. Because it is typical of their strong beliefs, it is translated here as literally as possible.

After I left here last time, I went to work for the government, and was sent down to work on the roads near the Umzimkulu River. There was a gang of us, about eighteen in all and we were camped in tents near a little stream where we got our water. This was a bad place, everyone knew it was bad. There were *abatakati* here, and the place therefore, had a bad name. We were afraid to leave our tents at night, and never did so when it could be avoided. The white boss was not afraid. White men do not need to be afraid. Either they are not in danger, or they have strong *muti* [medicine] which protects them, but *we* were very much afraid.

One night we ran out of water. It had been a hot day, and we were thirsty. None of us wanted to go out, but we all wanted water. The stream was only a short distance away, but the bush was thick and black. We wouldn't go, so we made the *umfana* [boy] take a bucket and go down for water. He had been gone only a little while when he screamed, and we heard the bucket drop. We shouted, but he did not answer, and we were afraid, but we went out with lights to find him. He was lying in the bush and was dead ["had fainted"; the Zulu words are the same], so we brought him back to the tent. He was breathing slowly, but did not know anything. We tried to get him to talk, but he did nothing but lie there. After a while he opened his eyes and was afraid but he could not talk. I sat with him, but he did nothing for a long time. Late at night he moved his arm and pointed to the pocket in his shirt. I did not understand at first, but after a while I knew he wanted me to look there. I did, and there was a little bundle of sticks, tied up with grass. He made me understand that he must drink it, and so I made a brew of them and he swallowed it. Then he got up and seemed well. Then he talked.

He said that he had just started into the bush when he saw an *umkovu* in front of him. He tried to move, but couldn't so he stood there. Then he fell down on his back, and the *umkovu* danced on his stomach. After it danced, it put something in his pocket, and said, "Drink that when you get home. I was a relative of yours once. I do for you what I can. I am ordered by another, but this I will do for you because you were related to me."

I do not know how the *umkovu* could talk. The boy got well, but I could never find the kind of sticks which he had in his pocket. Perhaps they do not grow in this country. It was lucky for the boy. He would have been dead now, or at least blind or dumb but for that medicine. I wish that I could find some more. I see that it was strong medicine.

Umditshwa was deeply convinced of the truth of the tale. It is probable that the boy imagined the episode, or else that he was subject to fits and hallucinations. The tremendous nervous strain of going alone into that dreaded place may have been sufficient to precipitate an attack, and I suspect that the persuasion used on the boy in the first place to go for water had not been very gentle.

Many people maintain that the native does not possess an imagination—a white man's superstition in itself. After seeing more than one girl who thought herself bewitched by a lover and became almost insane, it is clearly evident to me that the natives can imagine things vividly. Their folklore, superstitions, and beliefs are highly imaginative, and certainly would not be as interesting and wholly novel as they are if they emanated from unoriginal, unimaginative minds.

The Zulus credit animals with powers of magic, some good, some evil, and there are very few rare or odd organisms that do not fit into the native's idea of the universe as a place seething with occult forces controlled by wizards and human beings in league with them.

One of the most feared animals is a frequently seen bird that to the ordinary, uninitiated mortal looks harmless or even ridiculous. Up and down the streams that flow through wooded areas, a bird about the size of a crow stalks solemnly along the water's edge and fishes for whatever food he can find. There is nothing distinctive about the appearance of the hammerhead, hammerkop, umbrette, or *tegwane* (*Scopus umbretta bannermani*), as it is variously known. It is a brown somber bird, always looking rather untidy, never showing any particular vivacity or gaiety, and usually silent. To most people it is simply a drab-colored stork or heron, in any case a very dull bird, but to the Zulu it is an intimate of the wizards and is itself possessed of no mean powers of witchcraft.

As a matter of fact, the hammerhead is neither a stork nor a heron but belongs in a family by itself, with only one genus and two subspecies. It is most typically Ethiopian, and whenever one finds a well-watered country in Africa, one or the other of the subspecies will usually be present.

As to witchcraft, there are interesting explanations for at least some of the native beliefs. The Zulus have superstitions concerning most of the conspicuous birds and many other animals. Curiously enough, several other birds are the object of almost the same superstitions as the *tegwane*.

A Zulu boy would not dare to kill a *tegwane*. The natives' belief in this bird's evil powers is so strong that they are never known to harm it, and all the laws and police could not give it more adequate protection than is already afforded by superstition. If one of these birds should be killed, the natives believe that various dire consequences will follow. If the combination of circumstances is particularly unfortunate, violent storms will arise that will wash the hills into the valleys and sweep away houses. With better fortune, only a bad thunderstorm will result, lightning may strike the cattle or even some of the buildings of the slayer, but death will not be quite so widely meted out. With really good luck, disease may follow, but only the person so foolish as to touch the bird will suffer, and even he may recover.

With so many alternative possibilities, it is not surprising that within a few days or months after the killing of one of these birds, some of the milder forms of penalty will occur. If anyone becomes sick while the memory of the killing is still fresh in everyone's mind, no other explanation will be accepted by the Zulus.

The *tegwane* is supposed to be so potent for harm that merely walking near its nest is considered dangerous. Once, while making preparations to photograph a nest, I took along a native to cut brush and carry the cameras. On approaching the tree where the bird's imposing habitation rested in one of the larger branches, the native moved more and more slowly. At last, when almost under the tree, he balked. Nothing would induce him to come closer; even ridicule was insufficient to overcome his fears. When I started to climb the tree, he begged me not to risk my life. A snake would come out of the nest and bite. I assured him that nothing would happen. Then he said that if I touched the nest disease or some other misfortune would afflict me. I told him again that there was nothing to fear and climbed on toward the nest. At last, seeing that he could not prevent my actual contact with the nest, he stalked off out of sight around a bend in the river, alternately muttering or loudly shouting for others to witness, that since calamity would arrive he, at least, would be out of sight and therefore not responsible for what happened. In other words, he washed his hands of the entire affair.

The nest was a monstrous structure, twelve feet in circumference and composed of sticks, cornstalks, reeds, and various other odds and end. These bulky structures are so solidly made that they last for years, and along every stream where a pair of these birds habitually live, they are conspicuous objects.

This nest was fairly typical, but entirely unlike anything one would expect from the many tales. There was only one entrance, although according to hearsay, there should have been several. The nest is covered over with sticks and debris a foot or more thick, with a cavity forming a single chamber, although by local lore, it is supposed to be built like an apartment house with several rooms. There were eggs in the nest.

On my first visit the nest was new, although it looked old and disused. The birds build the structures of weathered material, and to the unpracticed eye they appear old even when new, in this respect resembling so many "genuine" antiques. Although the outer layers were simply composed of sticks dropped one on top of the other, jackstraw fashion, the vital part, the chamber itself, was solidly built. A man can stand on these nests and even walk around on them without danger of their collapsing.

The eggs are pure white when fresh, but a day or two after they have been laid the parents track mud and dirt over the clean surfaces, hence the eggs looked as though they might be too ancient to handle.

After a week or two the eggs hatched (incubation was in progress when I first found them), and the young hammerheads emerged as beautiful little bundles of soft gray down, the only time in their lives when they could maintain a claim to beauty. Except for two unusual circumstances, it would have been pleasant to watch the young birds develop. One was that, in order not to disturb the parents, the nest had to be repaired each time the young were taken out. The second was that soon after the young had hatched, the parents began to bring old bones, and rotting meat, and malodorous pieces of rawhide, and other unsavory objects that must remain nameless; it seemed as though the birds were bringing all the available filth from miles around and placing it on top of their nest until its odor resembled that of a small rural outbuilding.

It is difficult to say just why this condition, which I observed in three other nests, should be attractive to the hammerhead. It may simply be due to their habit of continually adding to the nest, even after incubation had commenced. Since this practice develops after the young have hatched and the parents have started to feed them, it may be that their simple reflexes are confused between debris-carrying and food-bringing, and the two impulses working in opposition may result in intermediate activities that induce them to bring material unfit for either purpose. It seems doubtful that they are scatophagous at any stage of development, although that remains as a remote possibility.

This unsanitary habit probably indirectly gave rise to the superstitious beliefs of the natives. It must be a fact that during epidemics of intestinal disease, any native disturbing the nest would be in danger of sickness, for the natives eat with their hands and do not feel it necessary to wash before meals. Besides, any cut or abrasion would be liable to infection. While I was examining the nest one day, an unexpected slip resulted in a mild "gravel rash" on my arm. I washed the cuts a few minutes later with clean river water and as soon as possible, in half an hour at the most, I painted the rashlike abrasions with iodine. In spite of the precautions, the slight scratches became inflamed, and for a week it looked as though the very painful infection might indeed become serious.

These are possible explanations of the superstitions, but there is another that might explain why the bird is held in awe by the natives. In the material forming the nest of almost any pair of these birds, one is certain to find an assortment of odds and ends of personal effects. At various times I have found incorporated into the structure pages of an old newspaper, a small gourd, socks, parts of cast-off underwear, other remnants of clothing, and, in one case, a small boy's *mutsha* or skin apron. The influence of civilization even on birds can thus be seen in many of the items. To a race that believes that the possession of a piece of personal property is a great aid or even a necessity for witchcraft, it is easy to imagine the state of mind resulting in an individual who sees his cast-off clothing carried away to the nest of one of these birds. Once a superstition has started, it is obvious that such a sight would clinch the case.

So firmly implanted is this superstition that even after the builders have left, few men will touch the nests, and the birds themselves are entirely safe from molestation. So strongly do the natives believe in the powers of the bird that they consider the sight of one flying over their houses an evil omen. A few days after my first visit to the nest, a specimen, presumably of the pair under observation, circled over our house and all natives agreed that danger was threatening and nothing could avert some calamity. As the weeks passed and nothing happened, the natives replied to the chaffing by saying that the hammerhead had done its best but that white man was immune from witchcraft.

Not only the hammerhead is feared but several other species of birds. The little wagtail that so confidingly builds its nest near houses and wanders about over the thatched roofs and in the yard is left strictly alone, one of the few small birds that is safe from molestation. The natives believe that an injury to the wagtail will bring swift and invariable retribution, most probably by fire that will destroy the dwellings of the family responsible for the injury.

Another bird that figures largely in their folklore and superstitions is the ground hornbill, *intsingisi* (*Bucorvus caffer*). The natives will never touch one of these birds except as a last resort during times of unusual drought. They believe that to kill one of them will bring rain. During the droughts, the death of one of these birds, especially if the body is thrown into a river, may produce heavy rains, a hazard worth risking when death by starvation may follow a drought. Apparently the less the ordinary possibility of rain, the more difficult it is to extract any even by magic. If there is a faint promise of precipitation, the natives will not resort to this method of rain making for fear that the natural rain supplemented by the torrents produced by magic, will combine to create a flood.

After I killed one of these birds near the Umzumbe River, the natives expectantly watched for rain, and although there had been no sign of rain, a slight drizzle set in not more than ten hours later. To the natives, this was proof positive that the birds were potential rain makers; only the fact that a white man, rather than a native, had killed the bird had prevented a serious flood. Strangely enough, my small gun boy, a lad about thirteen years old, had taken the body of the skinned ground hornbill and thrown it into the river in orthodox fashion, and about the same time another of these birds that I had wounded was found dead by another small boy and carefully thrown into the river. When the natives had seen me hunting these *intsingisi,* they pleaded with me to stop, saying that disaster would be sure to follow. The adults were decidedly afraid of the consequences. Probably the little boys were just as convinced, but one can trust a mischievous youngster of any race to experiment in the hope of excitement, even to the extent of doing what his parents would be afraid to do.

A relative of the ground hornbill, the trumpeter hornbill (*ikunata*), is considered just as potent a charm, but with entirely different effects. These birds are usually found feeding on the wild figs and other fruit in the bush along the higher hills and in ravines along the rivers. They are seldom seen near civilization and prefer wild and lonely places as their abode. Of all strange bird calls, the trumpeter hornbill's is among the most impressive.

One may be quietly hunting through thick bush. In the cliffs above, the pigeons may be gurgling hoarsely, and everywhere the small birds will be heard going about their various affairs, twittering and rustling, lending an air of sylvan domesticity and security to the scene. One soon may become beguiled by the peaceful sounds when suddenly, the wildest screams of brassy-timbred notes may blare out loudly, call after call, until the echoes from the cliffs peal out in every direction. It is a wild and blood-stirring sound even to one familiar with the bird that makes it, but when heard for the first time it seems utterly ghastly. A flock of these birds, when calling together, produce a marvelously weird clamor, and it may be that these shrill, discordant notes suggested the superstition that has endowed them with supposedly occult power.

After I had collected several of these hornbills, word went about, and for two or three days many young men, some of whom lived at great distances, came to ask for the gift of fat or the chance to buy it. Their own primitive weapons make it difficult for them to get the fat themselves. On my refusal to give until told the reason for their pressing desire, the men proffered what in some cases amounted to fabulous prices for a small vial full, but no explanations were forthcoming.

My young native gun boy (Mabulukwana; i.e., "Little Trousers," his real name) was anxious for his boss's reputation and good name, and after every request begged me not to sell or give away any of the precious fat; and he further warned that the carcasses should be burned to prevent any man from obtaining even a trace. I had noticed that all men who were so anxious for the fat were young bucks in their prime, gay and amorous men with no family ties, although because of polygamy and courtship customs, this latter fact need not have been more than a coincidence. When I pressed for an explanation of the strange desire for the fat, the gun boy was evasive, but a threat to give some away at last brought the answer.

Each man who had come for fat must have had at least one girl who was able to resist his advances, and each was after some of the *ikunata* fat which was held to have the power of breaking down this resistance—to win the girl's consent despite her lack of affection. For the men who had gathered around, however, consent by witchcraft was evidently as satisfactory as submission through affection, judging by their desire for this sex philter. To use this unique unguent successfully, a man would smear some of it on his hand and pay a visit to the stubborn girl. If threats and blandishments failed, the man was supposed to slap the reluctant "virgin" lightly and then go home and await results. Within a few hours at most, the girl must tear off her clothing and then, shrieking like an *ikunata* and waving her arms, the charm is said to drive her full speed to her bewitcher's house. If she fails to yield to the charm, she is doomed to pass the rest of her life wandering over the hills, wailing like a trumpeter hornbill. This form of seduction or magical rape seems abhorrent to many natives, and yet the men were unabashed by the presence of many spectators watching them trying to buy the magic fat.

The men were absolutely convinced that they could gain access to some coy girl's heart provided they could obtain the fat, and it was to prevent my being a party to such a scandalous proceeding that the boy pleaded so earnestly. Although a regular little demon himself, getting into every local fight on the faintest excuse and up to all kinds of mischief, he felt responsibility for his boss and had done his best to avoid an unwitting involvement in the unending promotion of sex activities in the neighborhood.

After learning the cause of his worry, I suggested that the fat from some other birds be passed off as *ikunata* fat and given to the native men. The boy would not listen, for said he, "If you do that, and some of the men use it, and happen to have mixed some *ikunata* fat with it, the charm may work. If they do not have any *ikunata* fat, and even though the charm fails, some other men may have the real thing, and his success will be laid at your door, for everyone will know that you have been giving it away. In this manner, even if unjustly, you will be accused of sharing in the seduction of some girl, and you should not run the risk."

This exhibition of wisdom coming from this little black imp was startling.

Mabulukwana was the only native, man or boy, willing to help with work around the hammerhead's nest. He climbed into it and uncovered the young without sign of fear. He held the young birds while I photographed them, and even carried some of the eggs home. Most remarkable of all, he would sit in the dense bush under the tree during the long hours while I took photographs and made observations from the blind. He could even sleep peacefully until needed to help take down the camera.

As the sun rises for us each day, the continent of Africa falls back into darkness, day in and day out in endless repetition, and each evening, as the night falls, the *usavolo* still whistles its melodious song as it has for thousands of years. So long as it still lives we have the reassurance that other small animals will also live on to perpetuate the special character of the African night.

The earth beneath our feet is much the same the world over, and each twenty-four hours we briefly share the same sun and the same moon, however different the moonlight may seem when serenaded by a coyote howl or a leopard grunt. Beyond the sky above and the common earth beneath our feet, the vast intercontinental differences we sense are biological, and they are the result of evolution and of isolation of the living things. It is the plant and animal life that produce the wondrous differences between continents, and they should be preserved, or all continents will converge in their similarities and leave for the future only monotony.

Even the superficial differences between man, differences of viewpoint, skin color, anatomy, and ways of living are products of evolution and isolation, and these too should be preserved to lend richness to the world we live in and comfort to those of different races. But on the African continent the years of superstition are numbered, even though ignorance still works through the fostering hand of the witch doctor-priests. Where reality is unknown imagination must serve to explain. As through all time, the translation of the unknown to the bulk of the people comes from their cultists, the witch or wizard-priests, who thrive best on the ignorance of their followers. But sophistication is spreading, fact is replacing fiction and folklore.

8

MAN

The fate of all living things hinges on man's need for natural resources and his attitude toward them. Excluding commercially
valuable species, plant or animal, most organisms become obnoxious to man when they interfere with expanding agriculture or
threaten already established farming activities. For this reason the
gradual but accelerating spread of man, his increase in numbers
over the world as a whole, continues to bring him into exploitative
or competitive contact with all untamed nature, spelling its im-
minent destruction.

Clearly, it has become impossible any longer to think of man's
future welfare, or even of "conservation" in its present concept of
hopeful hoarding, nor the future of any useful living thing without taking cognizance of the fact of man's animal capacity for
rapid and limitless multiplication. This similarity of man to healthy
rabbits liberated on an island without predators leads to the following discussion of matters that earlier in our history might not
seem to fall directly within the legitimate concern of a naturalist.

There is another reason why a naturalist must make brief excur-
sions into anthropological affairs. It is that whatever else man may
be, he is indubitably an organism, an animal, a mammal, a primate,
and hence responsive to environmental factors in a manner that
at the fundamental level is simply that of any animal organism.
Therefore his study is not reserved for anthropologists alone. Since

229

Homo sapiens is also an animal, he must be included as a member of the local fauna wherever he may be found.

Just where anthropology, sociology, economics, political science, and possibly religion, should seek to establish their boundaries will no doubt be attended to by each specialty and defended against trespass, but from the biologist's standpoint there can be no sharp boundaries other than those imposed by his own individual limitations and interests. The naturalist will of course have his difficulties in maintaining an unbiased approach while dealing with his own species; but this is not unnoted in the views of those claiming special rights in matters pertaining to Man.

With this excuse for attempting to understand man's past, present, and future relationship to his environment, that is, to his resources and his reactions to change in his resources, an effort has been made to look into the future of conservation and preservation of nature in the light of man's truly animal behavior.

THE AILING SOIL

The golden sunshine of winter for months on end falls on vast expanses of dry grass gleaming like ripe wheat. Its vibrant glow is accented in an ethereal quality by the lavender haze that veils remote distance, but it is also marred by black patches of ground where the annual grass-burning, for a time, leaves only a blanket of carbon and ash.

A person magically transported from exile in a distant land, back to his South African homeland, even blindfolded could guess the season of the year, because the fragrance of lightly commingled smoke and pure air would be his clue. These smudges of smoke against the pale blue sky are trade-marks of his homeland, originated by the extremely ancient practice of the native hunters and herdsmen and still employed by both whites and blacks, of setting fire to the drying grass each winter.

When populations were small and widely scattered, burning was spaced in time and topography and no harm was done to the soil or human welfare. In fact, it is possible that much of Africa's savanna country, its grasslands, and its scattered fire-resistant trees, are the emblem of a balance of nature created by burning and with such a many-century-old ecology that it now requires occasional burning to maintain the equilibrium in these ancient grazing and hunting grounds. It is possible of course that even in prehuman times lightning-started fires may have set the stage even before fire-using man arrived. In fact the statistical odds suggest this source of man's fires rather than a volcanic one. In either case there is in the ecology of annual herbs and perennial grass, and the bush and trees, a delicate balance of survival, and present conditions seem to be clear evidence of the continued need for occasional fire wherever man may wish to maintain the vegetational status quo.

On many farms wise use of burning even today improves or sustains the natural grazing capacity of the land, especially in the late winter months. Burning simulating the effect that presumably established the present ecology of grasslands is, at least for the time being, beneficial. Although at one time any burning was condemned as harmful in the long run, it was undeniably useful in producing a temporary emergency crop of fodder when the dominant grasses were otherwise too dry and indigestible to be useful. There is a miraculous regrowth of tender green and edible shoots that spring up after fires, even before the rains start, and so provide cattle and wild animals with grazing when otherwise there would be starvation.

Added to these first obvious results of burning was the relief brought about by the reduction of pests, insects, snakes, and rodents; however, their numbers diminished only temporarily and they soon bred back to their original abundance. In the light of afterthought and experience this rebirth of both good and bad should be understandable since the animals, like the vegetation, were adapted to the hazards of fire and would even be expected to depend upon it to maintain all those conditions to which they also were best adapted. With no burning, grasslands produce as highly specialized a habitat as does a forest in equivalent stage, and thus impose restrictions on survival, and require special adaptations as rigorous as in the better-known forest sequences.

The first condemnation of burning coincided with the observation that deterioration of the ranges and erosion of soil was seriously accelerated by removal of the vegetation, and that burning made the damage so extensive and serious that it was felt something should be done to preserve natural conditions especially for soil preservation.

As so frequently happens, knowledge and discernment were for a time insufficient to penetrate to the actual cause of these undesirable conditions. The synergistic effect of overgrazing plus burning became a major factor in reducing the number of desirable plants. Cattle, and of course large game animals, select the best and favorite forage, leaving only the undesirable to grow, mature, and seed, and so "weed species" eventually occupy the land with the woody and objectionable ecologic residue of unwanted plants. Balanced grazing and the intelligent, controlled use of fire to maintain the original ecology of the land, but especially the balanced grazing, appears to be the only process by which man can sustain maximum long-term natural production from the renewable resources. With wholly artificial forms of cultivation, range fertilization, introduction of foreign plants, or reseeding with native species after removal of native "weeds," time will produce a wholly man-made ecology.

Intelligent use of fire accompanied by controlled grazing in South Africa may be practical for the white man whose farms are still large enough to leave room for maneuver, but for the majority of the natives with their little fields and grazing grounds, no such maneuver is possible—theirs is already a hand-to-mouth existence. Any interruption in each year's productiveness would mean starvation.

For a time at least, South Africa will continue to bask in the winter sun and its hazy mellow distances will continue to display the rising columns of smoke that, since man's time began, have marked the regions of his abode. For a time this fleeting mark of man's presence will mark the grass- and bush-covered plains and hills—for a *time*, but at the most for only a very few generations.

In some of the more crowded native reserves, especially those that include steep drainage slopes—and these are very common —sheet erosion is widespread and so are the great gullies that are growing with each storm and that often supply no forage for cattle. Thus the overgrazing that brought on these gullies still further reduces the total amount of food available for goats, sheep, and cattle.

Obviously, these denuded gullies and the even more widespread areas of sheet erosion, suffer damage more rapidly than areas well protected by grass, and gullies tend to increase their size and capture ever more land. Once the unprotected top soil has been removed by heavy rainfall, regrowth of plants is retarded even without such competition as the numerous animals produce.

It was obvious to me as I revisited the reserves that while irremediable erosion is not yet universal, nonetheless it has occurred, and damage is becoming more irreparable as the years pass. With similar factors at work everywhere, damage to native-reserve lands is inevitable in almost every place where man is crowded. And land is uncrowded only where disease is still prevalent or was until within the past few decades.

Uncontrolled fires destroy the dead blades that ordinarily pad the earth's surface until the advent of the rainy season, and otherwise would add to the badly needed humus in the soil. The ash left after fires might be of limited value on nearly level ground, but even there it is so light that high winds blow it from the land, and the first freshet washes away whatever is left. Under natural conditions, the most important function of dying vegetation is soil protection and enrichment.

Not only is the ash washed away, but the ground is left naked of protection so that after the first rains and before the few undamaged seeds sprout, each little plant and tussock of grass becomes a tiny island that channels the flow over a narrowed space. Between these islands, the hard, bare soil, with no protecting admixture of humus, shows the minute scars caused by the rapidly flowing water.

To the effects of grass burning and an increasing native population, one must add the damage produced by an enforced shift from cattle to goats and sheep. The Zulus are fully aware of the differing ability of cattle as contrasted with the smaller stock animals to subsist on scanty food. They realize that cattle cannot do as well as previously. When their bovine animals find it ever more difficult to obtain sufficient food on the pauperized forage, the natives substitute sheep and goats, especially goats. Each increase in the numbers of these animals necessitates a still greater decrease in the number of cattle, hence the natives add to their holdings of goats. The change-over is accelerating with time, and the inevitable consequences are predictable. It is unnecessary to portray the details of damage done by heavy stocking with goats. The facts are already too familiar to too many countries.

To many white people the product of increasing impoverishment of the native means nothing except that more natives will be forced to work either on the white man's farms or in the cities. Too many shortsighted whites may even consider this result of deterioration of native lands desirable. The thoughtful Afrikaners, however, look beyond this ephemeral solution to labor-supply problems. As they see it, increasing impoverishment in rural areas and crowded labor markets will entail even more legislation for the protection of the poor whites, and hence deteriorating race relations.

Erosion will affect not only grazing land; the fertility of garden patches will also diminish and so will spring and stream flow. Any further reduction in the food and water supply is bound to become a serious matter. The natives will certainly not blame themselves for the many factors that have led to reduced food production, especially their multiplication in numbers which is the basic cause of the situation. The steps by which poverty has struck are too intricate for them to understand. They will believe only that segregation on poor land caused their troubles. Even when loss of productive land is exceedingly acute, and the natives are informed as to the remedies, most of them cannot, because of the nature of the terrain, take the matter in hand and alleviate their difficulties. The already low levels of nutrition, lowered energy, and the enormous effort required to protect even small areas of land would be sufficient to drive any race to despair.

That these conditions will not affect only the native blacks should be obvious. To their present grievances concerning decreasing fertility of lands will be added increasingly bitter rankling against the size and nature of their reserves when compared with the size, extent, and fertility of the white man's farms. The glaring contrast is inescapable.

Even within this decade it is obvious that the natives are discovering other weapons than those of violent warfare. Labor unions are becoming popular, and agitators of many kinds are finding friendly ears. It is unwise and useless for the whites to add fuel to the smoldering embers of dissatisfaction. Destruction of the arable land in the native locations will not only do this, but since it affects the native in his most susceptible spot, home and tribal life, it will also fan whatever other embers of trouble are present.

General strikes, vandalism, sabotage, incendiarism, are all weapons that the native can and may use. These methods of enforcing demands have been employed by many races against their fellow men, and it is difficult to believe that a subject race will not resort to similar acts when sufficiently hard pressed. The native, with his custom of not informing on other members of his tribe, will be exceptionally capable of carrying out violent acts as methods for obtaining his demands without detection by the authorities. These troubles of land sickness can have repercussions on the white man's status, and in the end they certainly will. Burning of white men's grazing lands, wattle forests, and sugar cane fields can seem wholly accidental, but will be very effective sabotage.

It is important for many reasons, therefore, to guard against erosion and loss of soil fertility and to check these developments as soon as possible, although the final requirement, no matter how intensely we may try to dodge the issue, will be birth control.

However we may explain the deterioration in living conditions of the natives, it is clear that had their numbers not increased they would still have an abundance of food and would be of little concern to the white man.

The fertile soil that survives the effects of overpopulation and erosion is further jeopardized by native habits—reckless cutting of building materials, overgrazing, nonuse of fertilizers, lack of crop rotation—and, of course, the ever-present goat.

Wherever I rode through the native-reserve lands, I could usually predict when I was approaching a native dwelling by the appearance of the bush. It becomes thinner and thinner as one nears the dwellings. The foliage is destroyed near the ground, and ground cover becomes almost nonexistent during the latter part of the winter when food is scarce. At the same time, a strong, pungent aroma, dominating all other odors in the vicinity, is usually noticeable, and sooner or later a flock of goats, led by a malevolent-eyed billy, turns to face the intruder. I often couldn't help but feel there was something devilish in the gaze of a goat. He looks more supercilious than a cat and has the same inscrutable stare that one finds in snakes with an elliptical pupil. To anyone aware of the mischief the goat is playing in accelerating devastation in these locations the innocent animal looks like Satan himself.

Originally the natives were in the habit of building their huts out of a framework of long, slender saplings cut from the nearest bush; over these frames was sewed a covering of thatch. The material for this sewing, the various parts used in the construction of the framework, and the binding matter for these parts were all taken from the bush. Thirty years ago the native was slowly abandoning his ancestral type of structure, but unfortunately he must still cut about the same amount of scarce material from the nearby bush or pay exorbitant prices for what to us are conventional and inexpensive materials. The long slender sticks and vines once used as *izintingo* (roof supports) are now woven into the walls of either a square or round dwelling, to provide the framework by which a mud plaster is supported. The roof is still of grass, but only rarely of the real thatch, which is now becoming scarce.

Unless one has had the good fortune to live or even occasionally to sleep in these huts (rondhovels), one cannot fully appreciate the changes that are taking place in the comfort, health, and general economy of the native.

The classical beehive huts with their clean, cow-dung-plastered floors (plastered once a week—or oftener in honor of a visiting white man) were warm in winter and cool in summer, and scarcely more smoky than some of the chimneyless, more modern type.

The grass huts were thick-walled, often with four to eight inches of genuine thatching composed of *isicunga* grass, each stem with a hollow, insulating center and dense seed heads that were loose enough to make good insulation. With such walls, heat transfer either in or out of the dwelling was so negligible that a small fire in winter kept everyone warm, and in the summer it was seldom uncomfortably hot.

The mud-walled huts by contrast, especially those with corrugated iron or flattened gasoline-tin roofs, conduct or absorb heat rapidly. They are cold in winter, become excessively hot during the daytime hours, and on summer nights the heated walls continue to stay warm into the early morning.

The gradual disappearance of building materials, together with the loss of tribal customs, encourages—perhaps compels—these changes. As one progresses into the heart of Zululand toward the Hluhluwe sanctuary, more and more of the classical beehive huts appear. Correlated with this change in construction and the tenacity with which the community adheres to the past, there is more grass and more bush to provide the building materials needed for the old-style structures. The rather lavish use of the classical materials is partly a result of the fact that residences (the kraals) are farther apart; that is, the population density is not so high, but this greater wealth of natural resources and lower population pressure can almost certainly be correlated with a higher disease rate, especially that of virulent malaria.

In conjunction with these activities is yet another that is potent for future trouble because it adds to the burden placed on any remaining bush. The natives clear the bush for gardens, and after burning all the brush and leaves, commence cultivating crops of maize and vegetables. Within a few years the fertility of the soil diminishes, as is quite natural under their method of cultivation in which neither fertilization nor crop rotation is practiced. Because of decreased grazing areas, as soon as the crops have been harvested, cattle and goats must be herded into the stubble where they eat every vestige of fodder and organic material that might add to the humus under ordinary conditions. When the fertility of the soil has been partially destroyed, more bush adjacent to the old gardens must be treated in a similar fashion.

Needless to say, there are other drains on the fertility of the bush in the vicinity of the kraals. Granaries and new huts for new generations must be built, and of course all the dead sticks are soon picked up by the women in their forays for firewood. When this method of collecting dry, dead fuel—called *teza*—fails, small trees are cut down, and limbs from the larger ones are hacked off, allowed to dry, and used as firewood. In clearing the land for gardens no forethought is exercised, and a tremendous amount of potential firewood is prematurely burned.

The firewood problem might be solved for a time by planting an acacia, now grown in large quantities on the white farms. Of all trees tried, only this species will supply both building materials and firewood for the native. The mere distribution of seed, however, will not serve the purpose. Even if the seeds are planted, the young trees will be certain to suffer neglect.

White men who have attempted to plant wattles in areas accessible to the native goats have had ample demonstration of their deadly effect on growing trees. The few white men who have been interested in supplying wattle seed and have helped start a grove, have been thoroughly convinced that the ordinary native is unable to appreciate future values. Even when encouraged and aided to plant trees, the native seldom does more than plant. After two or three years at the most, the grove is abandoned to the goats, and once more the native trusts to surreptitious efforts in appropriating the materials which he needs for building, usually at the expense of the white man's patch of bush or wattle plantation. It is therefore not always altruism that prompts the whites to supply neighboring natives with wattle seed free of charge.

One cannot blame the individual native for his behavior. There are probably many who under other social circumstances would be capable of looking to the future. Under the system of tribal, that is, native-reserve government, he does not own the land he is cultivating. He can be dispossessed and moved to another farm at almost any time. Although this does not happen often, it does happen, and unless the man is a favorite of the local chief, he is quite likely to be moved at about the time his improved property is becoming noticeably more desirable than that of his neighbors. I have had several occasions to talk with natives apprehended while stealing fruit and with other natives interested in putting a stop to such stealing. When asked why they do not plant fruit trees themselves, the answer is almost invariably that they do not consider it profitable to make their property noticeably better than that of their neighbors. Security of tenure is so uncertain that they feel it not only useless to better their condition, but also unwise. Doubtless many natives simply use this as an excuse, but there are others who know altogether too well that it is true.

When a man is moved from land that he has improved, he loses all benefit of work that has been expended on his place. He leaves a home for which he has formed an attachment. Probably he will have to build a new house and clear all the ground necessary for his gardens. Under these circumstances it is not fair to accuse all natives of laziness or lack of foresight, even if this may be true for most.

Not only does insecurity of tenure prevent a man from improving his property, but custom and superstition also enter into the situation; there is no force so compelling to a Zulu as custom, unless it is superstition.

A man who refuses to share fruit and wood—articles that have always been obtainable in the bush, growing wild and hence considered public property—is at once called stingy. He may wish to share a surplus with close friends or relatives, but if he shares with one he must share with all. To please his neighbors, even to avoid their displeasure, he must be ruinously generous. As soon as a man refuses to share everything that his neighbors believe he should share, he and his family become the focal point of envy and hate, and he is threatened with becoming the object of witchcraft. Life becomes fear-ridden and almost unendurable.

Consider our own country. A man living in the desert near a road puts down a well at considerable expense and trouble. The flow of water is not great. It is enough for himself and family and a little more. If that man sells the water, or refuses to give it to everyone who wants some, he is at once branded as a villain. This is comparable to the native's problem.

When a native has incurred his neighbor's hate, he soon finds that cattle are always getting into his crops, or that his boys get into fights when they attempt to herd their cattle on the common grazing land. His own animals get into other people's gardens or are injured in mysterious ways, and trouble multiplies in every direction. In court, all testimony goes against him; even the few friendly natives are not anxious to testify in his behalf. Almost invariably, sooner or later, threats of sorcery reach his ears, and he is soon blaming every misfortune on the witchcraft of his enemies.

Almost all natives are superstitious; even the educated ones, who profess disbelief, are only thinly veneered with enlightenment. Once they have been threatened with sorcery they are in constant dread of the day when the threat will be carried out, and it is not surprising that sickness, ill health, accident, failure of crops, or any other misfortune is attributed to a neighbor's hate and sorcery. Eventually they are convinced that living under such handicaps is not worth the effort. Either they move or yield to custom. Occasionally, of course, the bewitched victims retaliate by burning the suspected family in their grass huts, but the inevitable result of such retaliation is a dreaded court case. Naturally the justices who preside over court cases do not admit witchcraft as a sufficient excuse for burning the perpetrators of the sorcery, and the victims of witchcraft find themselves the victims of what they consider a rank injustice.

A more enlightened native may thus ordinarily be driven out, but there are exceptions. One well-known Umzumbe valley native was much more capable and industrious than his neighbors. The land assigned to him was ordinary farm land, but as he learned more about agriculture he began some rotation of crops, deeper plowing, better cultivation, and use of fertilizer for the impoverished soil.

When his land began to produce increasingly better crops and stood out in strong contrast to the scrubby fields of his neighbors, the local subchief and his retinue commenced proceedings to have him removed and started a campaign of hate. However, the native was intelligent—and at the same time distantly related to a powerful neighboring chief. More important still, he was one of the very few who were not superstitious. He refused to be put to flight, won all court cases brought against him, and in return instituted proceedings against the subchief. He won these cases also, and became one of the outstanding men of the district.

The few natives who own their land outright—the "freehold landowners"—are more fortunate, but those who understand the accomplishments and requirements that must be fulfilled before a native is allowed to purchase land, realize that such persons will always be exceedingly rare, and under any circumstances they do not come under the heading of reserve natives.

ANALYSIS AND EXTRAPOLATION

Demographic changes on the African continent have developed so rapidly in the past thirty years that practically every movie-goer, traveler, and hunter is still inclined to think of Africa as wilderness filled with roaming big game, virgin forests, and untilled plains still existing in their pristine, dawn-of-life conditions. There are many areas from which animal life, and vegetation as well, have been obliterated. Like cancerous growths these areas are spreading, the cells of this growth being new human individuals and new families, and there is no end in sight now that man has such effective control over his health.

It is true that much of the unhealthful equatorial areas—especially below the 3,000–4,000-feet level, as well as adjacent somewhat less tropical parts of the Continent—can still give the illusion of an unspoilable wilderness, and so do the remote desert and semi-desert parts, but the temperate areas, including the high central plains that comprise the heart of the Continent even in the equatorial belt, especially wherever soil fertility is good, no longer resemble the stereotyped concepts. In the healthful and fertile sections all has changed except in the forest and game reserves.

Possibly the most shocking phenomenon has been the speed of the destruction that has already taken place in the Union of South Africa, once probably the greatest big-game country that modern man has seen. That "progress," chiefly in terms of inevitable agricultural competition with big game, should have moved so fast in South Africa is almost inconceivable unless one has visited the country at widely separated intervals; but that damage should have reached the point where even the vegetation has suffered seriously seems utterly fantastic.

The extent of destruction to native plants even in the semiarid parts of the Cape region is scarcely known to Americans. The loss is particularly regrettable because the Cape, an area with a climate and terrain much like some parts of California, has one of the most diverse and interesting endemic floras of the world. It is a significant index of the inroads of man that, in this section of the Continent, sanctuaries and laws for the protection of native vegetation are even more necessary than in California. If such great damage has already occurred in the recent past and with a far smaller population than at present, then what of the long-term future of South Africa with its still rapidly growing population? It seems truly incredible that even here in this seemingly remote part of the globe man has exercised his destructive influence in so short a time. Inevitably one can only ask if man will ever come into nondestructive balance with his fellow living things. It is in the signs of impending and even greater human crowding and concomitant greater destruction that one sees cause for grave concern over the permanent preservation of African wildlife.

The mentioned changes are all man-made, and it is significant that directly or indirectly they have been made chiefly by highly civilized man only, hence chiefly within the past century.

Ancient man seems to have been in harmonious balance with his environment. He was an integral component of the "web of life," and even after the discovery of fire his numbers must have remained at a nondestructive level.

The rapid disappearance of wildlife in South Africa has caused a corresponding increase in the number of visitors to the national parks who come to see it while they can. Tourists are so numerous that automobile traffic en route to Kruger National Park has at times resulted in traffic jams reminiscent of those in the United

States, and housing accommodations within the parks have proved totally inadequate.

The presence of such multitudes of interested persons provides an unequaled opportunity for authorities to educate the general populace in the need for conservation and preservation. This chance is the only favorable result of the rising human population, and it could be used to maximum effect in expanding the numbers of the well-informed and appreciative, those who will be defenders of the reserves in time of need.

The problem of obtaining adequate funds with which to carry on the program of education, research, development of greater carrying capacity, for food and water, administration, and patrol are fully comparable to similar problems in the United States. Furthermore, crowds of people are everywhere much the same, whether in our U.S.A. (the United States of America) or the southern-hemisphere U.S.A. (the Union of South Africa). The ratios between the sensation seekers, the unruly, and the irresponsible on the one hand and those having more worthy interests in the parks also are practically the same.

On the whole, and despite limited budgets and correlated problems that are so familiar to us in the United States, increasing numbers of the general public in South Africa have become conscious of their reserves and are proud of them. The accomplishments of a small country of some 3,000,000 whites (and financially they are the only ones responsible for what has been done) justify their pride.

So far as the government is concerned, safety for the reserves seems to be permanently assured because of the steady growth of metropolitan populations and their political strength as opposed to the rural and often land-hungry segment. Despite the 1957 unsuccessful attempt to open the beautiful Mkuzi and Nduma reserves, the dollar value of preserving some of the Old Africa still available for display as a tourist attraction is becoming realized. Since it is the romantic and barbaric Africa that has by far the strongest attraction to tourists, it is clear that these remainders from the past must not be lost. It is significant that synthetic Zulu barbarism is becoming a drawing card at the Valley of the Thousand Hills in Natal, near Durban, just as the American Indians are at the Grand Canyon in Arizona.

But, synthetic or not, this sort of attraction is a harmless concession to tourist taste and should be added to all South African wildlife sanctuaries. The game guards and guides appropriately might be decked out in their classical African finery, with shield, spears, clubs, and monkey-tail or leopard-tail (*mutsha* and *betshu*) leather aprons. If to this could be added a native kraal as a going concern complete with all appurtenances, visitors as well as residents of the country could add to their knowledge of the past. Of all the native fauna, ancient *Homo sapiens africanus* was the dominant and most interesting feature, and for as long as possible he should be represented in the natural scene whether genuine, as a number still remain, or as an artificial but authentic replica in the long years to come.

But this ornamental reason for having the native present is not the only one. It is the custom for conservationists everywhere to evict the native *Homo* from all sanctuaries, but there is a legitimate question whether wholly natural sanctuaries, that is, permanently natural wilderness areas, can be sustained in perpetuity without including in the fauna the kind of man and human effects that participated in producing the precise conditions we seek to preserve. Without man we will drift to something wholly unnatural no matter how beautiful the result may be. Reintroduction of indigenous *Homo sapiens,* with freedom to pursue his ancient ways (other than uncontrolled proliferation) and to use fire as he did in the past, would seem to be as nearly scientific as we could get if we are interested in preserving nature and not in developing an arboretum and zoo. There are virtually no genuinely untouched "primitives" remaining today and if there were, they could not be kept purely primitive: the prejudices of civilized man would prohibit their free use of fire, hunting, habit of migration, interclan or intertribal wars, infanticide, and so forth. *Homo sapiens* as a natural organism in a sanctuary is probably no more attainable than *Homo* imprisoned next to gorillas and other anthropoids as a zoo specimen. But reintroduction should be tried nevertheless.

In some South African preserves, removal of the natives from among the other animals has probably also meant the loss of one of the most influential ecological elements and may explain, in part at least, why congregations of game now cause erosion, siltation of the ponds (pans), water shortages, subtle alteration of grazing conditions near water supplies, and various other vegetational and ecological changes. If this proves to be true, conservationists must devise substitutes for their most interesting animal, the one predator that has been barred from the reserves.

Even when primitive man was not solely a hunter (and he was often a herdsman with small cultivated areas to supplement food from his herds), he was a user of fire and a grass burner for hunting and clearing. His mere presence around water holes and good grazing made of him at least a stirrer of the wild animals, keeping them circulating to other grazing areas and watering places. His activities were certainly integrated in many ways with those of the animals surrounding him, and his omission from the sanctuaries may require conscious duplication of his influence. This probably will be as true for North America where the primitive Indian has vanished, as in Africa where the primitive black man is vanishing.

It is a curious commentary on our objectivity that we self-styled "humanitarian and civilized white men" inveigh against the African's poaching, his use of the admittedly cruel big-game noose (now made of steel cable filched from the mines), or the ancient game drives ending in pits, fire drives, and other habitual hunting practices, because if we wish to preserve genuinely natural conditions, cruelty and sentiments have no more validity than our references to the cruelty of the weasel, snake, hawk, or internal and external parasites. Our only concern should be to introduce no artificialities leading to any unwitting interference with nature.

If biologically minded anthropologists or zoölogists wish to place on record the important ecological activities of primitive man, especially in his relationships to big game, it will be necessary to start the studies at once. Too little opportunity remains to guarantee that even a reasonably complete record can be obtained.

Up to the present there have been no books or documentary films that have fully dramatized the intimate and many-faceted relationship of African man to his wilderness environment. The closest approach has been provided in written accounts and pictorial rendering of the Masai in their lion-killing exploits, with most of the emphasis placed on the dramatic appeal of the act. Studies of the lives of the tribesmen of wandering hunters—those who lead the most primitive existence—should be made as soon as possible or there will be no record of these details of their lives. Preliminary study and the making of elementary documentary films would require several years of work; in the meantime the natives who have been pushed from the game reserves have adopted other customs or have come to rely on the white man's intervention with modern weapons. These people are forgetting their skills and much of their wildlife lore as well as most of the many aptitudes that would be required if they are to serve in making an authentic documentary film of their ecological relationships to the wildlife around them.

For the long-range program of wildlife conservation, the exclusion of native blacks from participation in nature education and conservation, other than as game guards or guides, may prove to be a serious flaw. There are no accommodations for them in the reserves, and they are unwanted there except as servants. This exclusion of nearly all natives, including educators and potential leaders, is a poor indoctrination on the value of the reserves, for people who might some day have a hand in what becomes of the wildlife.

In a land of growing nationalism and mounting racial tensions, as in South Africa, the gravest danger looming in the future is that of a native revolt. Because the natives alone are not being educated in the value of the wildlife reserves, they will view the sanctuaries as they have in the past—only as a source of food and meat. If they were well armed with modern weapons, as they presumably would be if such a revolt took place with the connivance of an outside power, or if they merely employed their old customs of driving game by firing the grass and bush on a magnified scale comparable to their present large numbers, the slaughter and scattering of the animals would probably be so extensive that the fate of the game refuges might be sealed forever. Without extensive preëducation and indoctrination it is doubtful if even a single native leader would have the foresight, the necessary support, or sufficient control over his people, to maintain the reserves against the onslaught that would inevitably accompany a large-scale uprising.

Because the ultimate fate of the animal populations will depend so much on the solution of the human problems, particularly in the realms of food and of race relations, it is necessary for those who are interested in preservation to take cognizance of all factors, including the prospects for an indefinitely long continuance of the white man's rule.

An inescapable element in a realistic study of the entire subject must include that of the human populations, their cultural levels, their present food supplies, and, above all, their growth in numbers and the concomitant destruction of the soil.

The presently existing large sanctuaries and effective game laws make any genuine threat to South African wildlife appear as alarmist, except at some inconceivably remote time. Concern over future food supplies seems even less justified because to anyone flying over or driving past the immense areas of seemingly lightly used farm lands, it is difficult to believe that food could be scarce. One has to look closely at the productivity of the land itself—which depends on rainfall, chemical content of the soil, and, above all, the extent of the erosion—to understand what is happening.

Except for a few specialists in water and soil conservation and demography, no one seems to see the rapidity with which symptoms of danger are developing. The rate of change is most apparent to those who have been away from the country for several decades and are therefore confronted by a summation of what has transpired. This summation can be compared to seeing some elderly relative or friend after an interval of many years and being shocked by the cumulative changes, of which the permanent companion is unaware.

It should be reiterated that in the native reserve lands, comparable to our Indian reservations, one sees the most startling evidence of growing populations and the added effects of small annual damage. In these areas one finds a preview of the future. The transformation that has taken place between the early 1900's and the mid-'twenties and again between the mid-'twenties and mid-'fifties is most ominous. It is clear that deterioration of living conditions has accelerated markedly in the past few decades and that it results from increased crowding.

Conservationists think primarily of the damage a growing population will inflict on surrounding wildlife, including plants, but there is, of course, a more serious reflex effect that is affecting the human beings themselves.

Conspicuous deterioration in the native standards of living began comparatively recently. Even as late as approximately a century ago original conditions still prevailed and big game of many kinds was abundant—both a nuisance and a readily available supply of meat. Grazing wild animals and domestic cattle mingled freely and competed for food. The native cattle served chiefly as a source of tangible wealth, the counterpart of our own "ostentatious display," but they also provided the important and healthful *amasi*, fermented milk, and were used sparingly as sacrifices to the ancestral spirits. Because cattle were only rarely slaughtered for meat alone, it was inevitable that when they were freed from attack by predators they multiplied at a rate limited only by disease, and overgrazing became a threat checked chiefly by the rinderpest epidemic early in this century. With control of disease by tick-control devices, overgrazing again became a problem. Sacrificial slaughter was insufficient to check the multiplication of cattle, but when it did occur it meant protein for all of a kraal's occupants as well as for relatives and friends. The meat was not wholly wasted on the *amadhlozi* (ancestral spirits), and it is probable that propitiation of the spirits frequently coincided with meat hunger.

Soon after the arrival of the white man and the ensuing abolition of many ancient African customs that had kept death rates high, the native population and their herds of cattle started the steady increase that has continued throughout the succeeding years.

To comprehend the rapid deterioration of living conditions which has accompanied the increase in population it is sufficient to travel through Basutuland, the native-reserve lands in the Transkei, Xosa, or Pondo country, in Natal, and in some parts of Zululand proper and Swaziland. These all provide frightening examples of what happens sooner or later where conceptions are not restrained and where life-saving techniques of civilization are put into effect even if at very elementary levels.

Northeast of Natal and below the high, healthful inland country, in other words deeper into the tropical lowlands, population growth has been slowed by a higher death rate. As a result of a higher incidence of disease reducing their numbers, the human survivors have more space to live and to graze their cattle. Under such circumstances, wildlife has diminished more slowly.

In Natal, the Transkei, and parts of Zululand, conditions have deteriorated so rapidly that there are still many old men, the *ixegu*, who have vivid recollections of the days when there were large numbers of all the African wildlife. Some *ixegu* not only remember the swarms of game animals but they can date approximately certain events such as the time in their youth when they participated in killing the last elephant ever seen in their locality. This is notable because the elephant **was** one of the first large animals to disappear.

These old men love to reminisce over their beer pots, but their memories are fading fast. In most native reserves these men tell of the days when the Zulu *impi* (armies) ranged abroad ravishing the country and keeping populations down. In those days the bush was abundant in the valleys, and between the bush and on the highlands, *etafeni*, there were miles of tall grasses rippling in the breeze and swarming with game. There was more than enough grazing for the wild and domesticated animals, and the ranges were never overgrazed. At that time, the small garden patches nestled in folds of the hills where the soil was rich and water abundant, and there was always room to move to another clearing. There was always more than enough unspoiled land to clear and plant. Of equal importance, the abandoned land could rest and recuperate indefinitely, for centuries if need be.

After the arrival of the white man life became gratifyingly more secure, but gradually, though within the lifetime of the old men, room to move about diminished, grazing became more restricted, and game disappeared rapidly except for the little duikers and reedbuck in the fields and the bushbuck and tiny blue bush duiker in the dense cover of the bush. But as the years went by, the human population continued to grow, grazing became poorer, and the garden soil poorer still. Trees and shrubs and the birds and animal life vanished from the most hard-pressed areas. In native lands the garden patches kept spreading until boundaries merged with other boundaries, and also the small animals, duiker and rabbits, and in many places even the cane rats, were obliterated. None of these vanished living things moved to better, safer areas: they just seemingly disappeared from the scene. Soon there will be no time nor place where the land can regenerate. Neither game laws nor rigorous adherence to protective laws could have saved the wildlife unless human beings had been prohibited from reproduction or from earning their living from the soil.

Today the old men huddle around the feeble little fires in their huts and think of the past. There is no more wood to cut for the fires, there is no longer any good thatch for building the old warm huts. The fires must grow smaller by the year, and as mud walls and thinner roofs replace the ancient beehive huts of thick thatch, even the best of large fires would be inadequate for the bones of young and old. In many places the people must sit ever closer to the smouldering fire, now sustained on dry cow dung with its bitter acrid smoke that has replaced wood smoke. If firewood is to be obtained it must be bought to an ever-increasing degree from the white man, and very few natives can afford to pay the price. With the dwindling grasslands, cattle also must go, and goats will take their place, but what then of cooking and warming fires? It is impractical to attempt to glean for fuel the minute pellets of compacted cellulose that drop along the goat trails. Besides, goats destroy the bush that might recapture abandoned land. Dependence on goats seems to herald an end to the supply of free fuel.

As man's numbers have increased through the decades, resources have disappeared: the bush, the vitality needed grasses, the necessary cattle, the firewood, and building materials. The very soil of the hillsides is going, and the once-favored bottom lands are being buried in sterile sand or gravel.

Even on the generous lands of the white man many changes are taking place, but they are generally not so obvious. Where the white owner does not divide and subdivide his land among his offspring, the holdings remain large enough so that with care the natural resources need not be overtaxed and destroyed. Except for avarice or overambition, the land could last indefinitely. But even though the sons and grandsons move to the cities and to the factories, they must still be fed, and thus they are continued dependents on the old estates: whether on farms or in the factory, the land is still their only source of food. Wherever farms must be subdivided among each generation's progeny, the fragmentation soon leaves inadequate land to support new generations. It must be overworked and it will deteriorate. Successive generations must be fed and the per-capita share shrinks.

There are also changes that seem harmful only to some and beneficial to other species of wildlife. In the sugar-cane belt, for instance, almost all land has been planted to that tall grass, and except for a few accidentally adapted or adaptable species— among them bushbucks and cane rats—most of the original endemic fauna and flora have been exterminated outside some isolated patches or thin strips along the river cliffs. On the higher hills, wattle groves or other afforestation projects displace the natural wildlife, but at least there and on the cane farms, the soil is held intact or even improved. Under other types of farming, however, erosion is apt to take place, but it is frequently only in an incipient condition as compared to the native reserves. On some well-managed farms there has been scarcely any damage, and the smaller representatives of the ancient fauna still flourish. Below these happy lands, the streams too show little damage. But pressure for more intense cultivation is in the offing for as the population grows, the sons of the landless, the dispossessed, must be fed; and the natives too, must have a share, or they will starve or revolt. As the need for food increases, even these lands may be forced to yield ever-greater crops, and the warning is clear that with too much abuse, the soil and the climate may sabotage man's plans. Whether he wishes to or not, in time the white man must assume the burden of feeding most of the black men, those from whom all natural sources of sustenance are being withdrawn by overpopulation and its sequel. The native will have no other source of materials or food than those passed down from his white ruler (unless he keeps his numbers down to what his land can feed) or he will become an industrial peasant. The white man is subject to the same rules, but he has learned how to defer for a time the inevitable penalty by modern means of extracting other kinds of wealth from the soil, that is, by the use of its finite and nonrenewable resources.

In South Africa the agrarian way of life among the already hardpressed natives is about to come to an end for all but a small percentage of their numbers (in 1956 reported to be about a third of the total). Lost to memory will be the barbarian, primitive, and romantic, except in synthetic replicas or the reports of a few interested and understanding scientists and novelists. Further generations of travelers, and South Africans too, must depend on these accounts and on the all too few and inadequate documentary pictures.

It is usually dangerous to extrapolate in biological or human affairs, but in South Africa the picture is so clear that, excluding some unforeseen check to population increase, it would be foolish to reject the implications for the future. Other generations will follow, and their welfare is our concern also. For the long times ahead, the African native reserves teach the lesson that unless we curb our increase, nature's penalties may be applicable not only to that remote continent, but elsewhere as well.

There is ample evidence that there is a limit to the total numbers of every living species—a limit that is determined by its resources. For all kinds of nonhuman organisms, the limit is enforced by famine, disease, or other death-inflicting components of environmental pressure. Only a high death rate can limit the ultimate result of a high reproductive rate, and nature is wholly ruthless in imposing its rule on any avaricious species that outbreeds its renewable-resource income. Only man can choose to limit consciously his reproductive capacity and thus avoid the indifferently cruel penalties that nature would exact.

There is little evidence to suggest that primitive man, especially agrarian man, will limit his population growth by humane and conscious voluntary means. Abortion and infanticide, as well as destruction of the aged and uselessly infirm, have been widely practiced but these can scarcely be considered as humane by modern concepts. There is no evidence that man possesses any innate, automatic, fertility-modifying safeguards against unconsciously and witlessly outbreeding his resources. So long as this situation prevails only the occurrence of famine, pestilence, bloody revolutions, or war will check his growth until he learns the obvious fact that even for scientific man there is a limit to natural resources. Except for real resource expansion in agriculture and allied fields, traditional science seems mainly to have achieved knowledge of mineral and fossil energy resources, and simultaneously to have discovered how to exploit them at an ever-increasing rate. For already overpopulated agrarian societies agricultural discoveries and western techniques have so far served only to increase the numbers of the people rather than the quality of living.

Scientific man, who now depends on many finite and nonrenewable resources (though in the ultimate sense he figuratively eats even these) as well as on food, might learn the simple lesson, that half the rate of consumption will double the life of a finite supply. Experience may demonstrate that this doubled time will be necessary to provide the basic information that he will need to extend his welfare to his grandchildren or even, for our youngest people, to their children.

There are many reasons why the white man, with his infinitely more complex dependence on resources, finds it impossible to realize his relationships between resources and population size. He persistently confuses a high rate of extraction and exploitation with abundance, rather than seeing in it exhaustion and depletion of nonrenewable resources. Our ever higher standard of living for ever larger numbers of consumers represents capital expenditure rather than subsistence on interest. Of the future of fission and fusion energy and other new resources there is hope, but we still have a long way to go before a final and firm answer is possible.

In South Africa there is confusion along these lines of thought, and failure to see in the modern and rapidly industrializing and overpopulating world the probability that in time food will be consumed locally everywhere and that there will be no exportable surpluses to be exchanged for manufactured goods. Their failure to see this possibility encourages them to emulate older populous countries and to place too great reliance on industrialization to feed ever-larger numbers of people, instead of keeping agriculture and food production ahead of industrialization and, putting first things first, thereby balance food, population, and industry. Rather than follow this cautious, safe procedure they are encouraging an increase of the white population (a device designed to ensure military and police power) and providing clinics, hospitals, and health instruction for the natives. This concern for the natives is primarily a result of economics and the labor supply, and partly due to the fact that although segregation of humans can be practiced, their diseases cannot be similarly segregated. There is a rather small segment of the political white population which supports health measures for humanitarian reasons alone.

It is far simpler to predict the ultimate consequences of these policies for unlimited population growth than to suggest an approximate time for their fulfillment and catastrophe.

Naturalists are often inclined to avoid concern in human affairs, even those that are illuminated by the picture of overpopulation in all other species, but even the escapist naturalist must sooner or later take cognizance of these matters in human terms. With or without humanitarian impulses he eventually begins to realize that it is only by solving the human problems that there can be any hope of preserving the things he loves—even of minute samples of the world's characteristic flora and fauna—and keep them flourishing down through the ages in an unaltered state.

The mental climate developing in the native population is inevitably grafted on older views concerning game animals. It is difficult, then, to teach the native the values of a wildlife sanctuary and the benefits of limitations on his population size.

Historically, the native has looked on game as his meat, his *nyama*, well symbolized by his collective word for all game, *nyamazane* (meat supply). When in due time he becomes seriously hungry and his children are starving, when his fields cannot produce enough grain, nor his pastures keep his cattle, how can he be taught the meaning of a sanctuary swarming with food, from which he is excluded by the white man's laws? How can any white South African justify to the natives the preservation of areas where the antelope and rhinoceros grow fat while the black man and his family starve?

But even before justifying the sanctuaries, there will be need for much additional explaining. Why, for instance, have the original owners of these lands been forced to live in slums or commute from their wretched but beloved homes in the reserves to work in the city? Why, also, should there be the enormous and seemingly unused farms of the white man lying right beside their deteriorating reserves? Above all, it is difficult to explain the existing relations between white and black.

Unfortunately, the situation in South Africa is rapidly approaching the point where fear and hatred will go so far that understanding can never be achieved. Yet the fate of the innocent wildlife ultimately must rest on human welfare and the solution of race relations. Man's own future also demands that these problems be solved unless either a new type of slavery—complete subjugation of the black people, setting precedents for other areas—or revolt and war will result. All parties wish for peace even despite their present difficulties. Furthermore, it is possible that the status quo might be preserved under static conditions, but the population is not static: it is pressing remorselessly upward while the hillside lands are washing away.

Any outside comments on the South African problem must be based on an understanding of many factors. It should also be remembered that the situation there is distinct from that found anywhere else on the Continent, the nearest approach to a similar condition being in Kenya. This dissimilarity is based on many factors, including the long history of white occupation in the South, and a resulting indigenous white population having no other homeland nor outside ties. Also, the original occupation of the Continent took place at the same time that we, in the United States, were dealing death blows to the Indians of North America, and when conquest, colonization, and slavery were still morally acceptable to both the conquerors and the conquered.

In the United States there is an almost universal tendency to base conclusions on our own experience with race problems, but, except to a very limited degree, this is not a valid procedure. Even in our Southern states there are few conditions that are helpful in making comparisons and drawing conclusions. In order to understand South Africa, it must be taken as it is and where it is, and not out of context with its geography, history, people, and social attitudes.

Americans are prone to oversimplified thinking and to utter *ex cathedra* suggestions for a solution to South Africa's problems. The almost automatic suggestion, of giving more land to the native because of his need, in effect means that because of rapid increase of the native population, land should be withdrawn from the white man and added to the territory of the prolific black man. This would be equity in a sense, equity based on birth rate. But it would be an endless process, and therefore it would mean a race, or a war, by means of the cradle. It would be scarcely fair to the babies.

Another seemingly logical, often-suggested solution is equality for all with no racial or color bars to delimit privilege. This would be justice of the highest order and would require mutual trust and confidence between the races, and a willing abdication by the white man of his position of power. Theoretically this might be achievable; practically, however, the feelings of the white man preclude it for a long time to come. The mutual climate for so drastic a change must be developed in spite of a history replete with war, treachery, and bad faith between these people as to common cultural and economic thinking and habits. Only a minuscule few at the apex of culture and the lowest dregs of both societies have anything in common, whereas it is the masses between these extremes who must have some common sympathies and understanding before mutual faith and safety for both sides can be achieved. Above all, there would still remain competition via the cradle for the ever-larger economic share per group, but not per individual, of the food of the country. Yet, already basic maize and wheat production is inadequate for the needs of the people, so that except for occasional extraordinarily good years these foods must be imported.

If there were time to educate, indoctrinate, and change the feelings of such diverse people, there would be reason for hope of resolving these tensions, but the rapid population increase and attendant pressures on resources do not provide time in which to operate. It has been said by a church that should know whereof it speaks, that it takes three generations to produce a genuine adherent of the faith. It seems impossible that in even less than three generations of population growth, pressures will not build beyond the explosive point.

Voluntary population control by either white or black peoples, or both, seems impossible of achievement within the requisite amount of time. Although unacceptable to the rural Afrikaner, and to most other people as well, for theoretical purposes one might reflect on the results of attempting to impose a unilateral population limitation! To induce the native to limit his population, with or without white participation in the program, seems impossible because of suspicions such a program would engender as well as because of the value of daughters in terms of *lobola*, the marriage dowry, and the earnings of a son that must be turned over to the father.

As for political stress between the people of Dutch and British extraction, it might be expected that there would be sufficient mutual interest resulting from a common source of danger (plus their common European heritage) to achieve rapidly a high degree of coöperation, but there is none. On the other hand, the dark-skinned people, for economic and social reasons, might be expected to achieve harmonious relationships among themselves, and thus still more effectively than heretofore split the South African population into two clearly demarcated opposing forces. However, even today there is seemingly little or no effective cohesion between the various races and cultures represented among the dark-skinned. In time, more misery and more suppression will unite the subject groups. It is even possible that despite its revelations of ruthless self-interest, communism may be the catalyst to bring about this fusion of tribal and racial desire.

South Africa is thus cursed with a man-made situation fraught with much tension and very grave dangers, yet no one seems to recognize that there is one basic and inexorable threat common to all, irrespective of race, color, or creed—the lack of adequate future subsistence.

As a naturalist, I can come to only one conclusion, which is that no rational solution can be achieved in time to avert disaster to both wildlife and man himself. Temporarily, the prosperity of the white man in terms of gold mining and the new uranium resources can be applied to the economy of the country and may serve to postpone but not escape the consequences of overpopulation. At present rates of consumption, coal should suffice for seventy-five years, but expanding consumption and population increase will accelerate. Population growth is exponential—progressing by doubling and redoubling. If energy from the rich uranium deposits can be used to add to the food-producing capacity of the country, the people may have sufficient time to work out their destiny, but the dreams of power from fissionable elements are still to be realized, and time is growing short. For the all-time fate of the land, its wildlife, and its peoples, it can only be hoped that conflict between the races can be averted. It seems probable that only a biological solution will be achieved, survival through force and ruthlessness.

Whenever we think of the future of animal and plant conservation we assume that if only the boundaries of present reserves are maintained inviolate, nature will thereafter persist uninfluenced by man; that in principle we only need to guard the boundaries from agricultural or industrial encroachment, and the woods and fields from invasion by poachers. In a sense this is theoretically true but in actuality it is, of course, false because we establish a boundary beyond which neither fauna nor flora can find their way, and into which no genetic material can flow. We are isolating a section of continental forms of life in an island from which there is no escape, and into which there is presumably no penetration by surrounding species. Biologists are well aware that neither populations nor associations of organisms exist in a passive condition, but are, on the contrary, exceedingly dynamic. These factors militate against preservation in perpetuity anywhere, but admittedly they operate at such a slow rate that they are invisible at any time or place. This is the best we can do.

The age of subhuman mammals has gone forever over most of the world. It is only here and there, scattered over the civilized world or in areas under the control of civilized man, that tiny dots of land, less than flyspecks on the largest global maps, represent the sanctuaries where man has seen fit to permit his lowlier mammalian kin to survive under more or less natural conditions. Outside of these, wild mammals are vanishing so rapidly that if changes continue at even the present rate freedom for the larger species to wander as they once did may not extend beyond the next three or four decades. Throughout Asia and parts of Africa even less time may see the virtual end of the more vulnerable, rarer, edible, and nuisance species.

Preservation of the unique African wildlife, both plant and animal, should be the concern of every civilized nation. These relics from the past are fully as valuable and vulnerable to social or military upheaval as art treasures, libraries, or even notable examples of architecture.

In time and with sufficient education all races of man might come to understand the importance of preserving these unduplicatable assets. Multiracial knowledge is necessary, because it is only through widespread appreciation among people as a whole, reinforced by their governments, that wildlife reserves anywhere can be kept in perpetuity. With the passing of each generation the values that are inherent in these reserves undoubtedly will become more easily comprehended simply because of the destruction of other faunas and natural conditions elsewhere, but in the meantime there is grave danger of their being irreparably damaged or even totally destroyed.

Carefully regulated use of our wildlife resources (controlled shooting or trapping) can retard the process of extermination, but the most lethal and remorseless weapons directed at our heritage will be the plough, the hoe, the ax, and the bushknife, which are even now relentlessly gnawing and destroying our resources, driven by the growing pressures of human fertility and falling death rate. These forces are digging out the land from under the very feet of the animals and ruthlessly chopping away their cover to leave them naked, exposed, and starving.

WITHIN THE LIFETIME OF A MAN: RÉSUMÉ

For uncounted centuries the tall grasses of the South African valleys—the Umzumbe, the Umzumkulu, the Umkomasi, the Umvoti, and all the others—had reached shoulder height, and at the end of the rainy season the long-horned native cattle were almost buried in growing fodder. Still later, Zulu invaders, as sleek as their cattle, leisurely tended their herds, and their small but adequate gardens of milo maize, "taro," squash, beans, and ground nuts, but most important, maize. From the product of their grain crops the warm, leisurely months of winter were filled with feasting and the revelry of beer drinks.

Except for occasional but recurrent murderous raids and petty wars initiated by their Zulu brethren to the northeast, life was outwardly peaceful and contented, rich in material things needed to gladden the hearts of these people. Nonetheless, fear remained an invisible canker in this Eden: There were senseless fears based on superstitions, and real fears of the death that could be meted out to anyone who might be accused of witchcraft. There was the ever-present fear for the death of loved ones at the hands of sorcerers. There was also the fear of old age and infirmity when one lost his usefulness and would be turned out into the bush to die.

Death as infanticide, parricide, murder, accident of the hunt or in the fields, and incessant wars and raids, kept the population small over the centuries.

Beyond these dangers pervading the valleys even then, was the fear that in prolonged droughts, famine and disease might decimate the families. Nonetheless the inhabitants of the valleys had less to fear than others on less fertile soil.

Despite their half-buried fears the people had adjusted to a satisfying routine of life. There was robust enjoyment of the delights of living, a minimum of work for the men, and now and then savage interclan fights to relieve the monotony of rich existence.

In the passing centuries before the coming of the white man the population of the valleys remained essentially constant. In spite of polygamy and ardent participation in the pleasures and profits of procreation (daughters were worth ten head of cattle), the population failed to grow appreciably. The multitudinous vicissitudes of primitive life kept death rates high, balancing the high birth rate, and there was therefore always enough land and to spare for the survivors. Freedom from want prevailed except in times of unusual drought, and even then the streams and springs and meat from wild animals never failed. The only unhappiness stemmed from the natives' bondage to their superstitions and customs, and the selfsame fears and consequent practices of their neighboring clans and tribes.

In time strange rumors began to seep into the valleys with increasing frequency. Stories of tall, quaintly clothed and fearsomely bearded and armed white men spread to the remotest kraal. There were even stories that these few strangers had inflicted defeats on some of the most feared tribes. Eventually these barbarian natives heard the sounds of guns for the first time. It was not long until hunters, traders, prospectors, and missionaries became actual visitors within the confines of the valleys.

These white men were an odd and confusing breed to the natives. There were many would-be despoilers of the land and exploiters of the inhabitants, but among them were also educators, philanthropists, government agents, and the most strange missionaries—a breed concerned only with people's thoughts, and superstitions, and nakedness.

It was not surprising that the whites were accepted reluctantly, and that none was trusted, including the missionaries who spoke of strange codes of comportment and of strange gods, yet dis-

agreed with each other as to the kind of god they served. Their
exhortations pointed only to a seemingly harsher way of life, and
converts were few. Nevertheless, slowly and reluctantly, new ways
were adopted by the Zulus, and fewer people died from the effects
of superstition. Homicide and infanticide diminished. Deaths from
feuds, raiding, and war ended; gradually the high birth rate gained
over deaths, populations grew, and each new home was marked by
the spiraling wood smoke that rose from new clearings being made
for family homes and gardens.

Through the early years the task of the missionaries was dis-
couraging, and progress among the natives was slow and plodding.
These pagan people were well fed, wives and cattle were plentiful,
graceful, and fruitful. Although the death rate among infants was
still high, 50 per cent or more, each year greater numbers survived through the benefits of mission instruction in feeding and
sanitation. Those that survived developed immunity to most lesser
ills and grew to be healthy barbarians—externally, at least,
replicas of their parents.

The natural balance between the forces of life and death which
had existed before the advent of the missionaries was gradually
tilted in favor of life, and populations began to expand. With ever-
larger numbers of families engaged in cultivating virgin land,
the forests melted under the attack of bush knife, ax, and hoe,
and ultimately the plow. On the higher, somewhat steeper, slopes
of the valley, the fodder grasses, almost before they could seed
themselves were disappearing into the mouths of cattle, and tall
thatch grass and young sapling trees, their essential building materials, vanished from bush and fields. Both cattle and gardens
commenced an ominous encroachment higher onto the surrounding hills and onto ever steeper slopes.

Where the Zulus had once been fat and sleek, their cattle numerous and with ample milk for calves and humans, now the rising
tide of consumers was beginning to outpace the natural productiveness of the land. None could see the relationship between
health, security, population increase and the diminishing abundance, which every household was feeling to some degree. There
was no understanding, only envy of the few whose lands could

still yield a moderate crop. Of those who spoke of the good old days none ever mentioned that in those physically comfortable years, only a half or a tenth as many people were forced to depend on the same acreage, nor that the extra space and wealth, the good land for crops, the grazing for cattle, the clear springs, the surplus of building materials—in fact all material blessings of those days—were bought at the price of a high death rate.

For the inhabitants of the valleys, hardships mounted steadily over the years. Even the richest fields were starveling and yellowed by surface erosion or burial under sterile sand from higher slopes. Famine had come closer, had in fact become a constant threat, and in bad years starvation could be averted only by transporting food into the valleys, always at a price that led to debtors' troubles.

The years, the decades passed, the life-saving benefits of science were taught in schools, missions, and churches, but nowhere was it ever mentioned that man and his numbers were dependent on a balanced population. Steadily more children grew into reproducing adults, and as the swarms of individuals multiplied, life for each grew more bleak. There is nothing to show that the happiness—if it could be called that—of the many who now survived was in any way better than the previous happiness of the few, even when it was purchased by death. For the Zulus of the valleys life had become very difficult, and, as Patton has written in *Cry the Beloved Country*, happiness may be "beyond the stars."

And so to a recent year, when the rains of summer had come to the Umzumbe valley at last. At least for another year the unspeakable weariness of waiting for water that did not come, the fear of too-long-delayed planting, and the sight of drying and ever more barren grazing grounds was over. All through the valley, for weeks so silent and oppressed by fear, relief expressed itself in frequent singing and the strange falsetto calling between distant clusters of huts.

There was joyful talk of the first results of rain, the freshening grasses and increasing yields of milk to be fermented into *amasi* for both old and young. There was keen anticipation of the first tender greens of *tsinkunchani* and other herbs that would soon provide a welcome and health-giving change from the monotony of corn-meal pap, for months the sole but vitamin-deficient article of diet. New hope livened the faces and quickened the step of all. The mere smell of fresh-wet earth and dampened vegetation overwhelmed the senses with new joy and spoke of the pleasures of plowing the moist fields and planting the painfully hoarded seed grain.

But the storm increased in intensity, thunder echoed from the cliffs around the valley and in between the periods of rumbling din, from every *donga*—steep-sided gulley—came the sound of trickling water. Even from the open fields sounded a steady whisper of falling rain. On the now barren surface of the ground, trickle joined trickle, united to streamlets, pouring into the *donga* to swell each stream and river. Each pounding drop of rain loosened more grains of soil, and the rapidly increasing flow bore the lighter particles, the scanty humus and silt, to the streams and thence to the river, where the speed and power of the water shifted first the sand and gravel, then stones, then rocks, then boulders in a steadily rising flood.

At the storm's end when the natives peered out of the doorways of their huts, the air was filled with the roar of the rivers and streams. The acrid stench of a river in flood was sickening. Overnight the reed-grown seep of clear water had turned to a raging torrent of swirling mud, a hundred yards wide, inundating and tearing away some of the best remaining lowland's garden soil.

Here and there on the river surface paraded the procession of the dead—chickens, goats, calves, even steers, and now and then, but rarely, a human body—all plunging and swirling toward the distant sea. When rain falls torrentially and when the natural cover of the soil has been removed, fear and sadness, having waned and almost disappeared, return once more to this and all the other valleys, but in a new guise. In every hut there is a dawning realization that under the beating rains and grinding torrents some fertility is lost, and that "rocks are sprouting" where the Zulus plant their fields. But there is still no comprehension of original causes of their plight.

This is the new pattern of the generations in this valley and symbol of multitudes of other valleys. Each year there is less fertile bottom land. Each year sheet erosion is accelerating on the sloping plowed land and on the dehydrated grazing grounds above the fields. Each year the cry rises louder that they plant the land and that the rocks grow in place of crops. Year by year less water penetrates the soil to keep the streams and rivers flowing. Even the remaining springs are drying. Now the fields are increasingly dependent on greater frequency of rains throughout the growing season. Each year there is fear that rain may fail to come, or be so spaced that crops may die out in the intervals between the rains, or that if they come at all, they will come as floods. The margin of survival for their crops, their stock, and themselves is growing more precarious by the year and by the generation. Dependence for survival is no longer safely placed on local crops— dependence on city jobs, and the migration to the city for subsistence grows each year. Some fields are growing thorn, and in time the land might rehabilitate itself, if left alone.

What is true of the valleys of Africa is also true of Central and South America, and many parts of the United States. As populations grow and more sloping lands—more acres vulnerable to damage under cultivation—come under the plow, it will be the pattern of the world. The lands suffer from erosion, but not so much from the flood of water alone, for that has always come, as from the flood of unchecked human multiplication and from the cry for more food, ever more food, and intensive farming where there should be no plows at all. And yet in South Africa as from the rest of the world, there rises the cry for more medicine, more life saving. At the same time, the flood of procreation goes unchecked. Almost nowhere is there a call for fewer people, for less destruction of diminishing resources, for greater comfort and security of generations still to come, decade by decade, and century after century.

There is no longer any chance of escape to a neighboring valley —there the process would only be repeated, for there the lands are already occupied, and there is no flight to the higher plains, for there also the land is occupied. There is no escape to other lands across the seas, for there also are people with insufficient food. For the time, there is a seeming chance for life by flight to the cities, but cities must be fed from the land, and at best the fugitive merely fills the ranks of unskilled labor, and there is no surcease from insecurity.

South Africa is not alone in its dilemma. It is a stage on which is being played the drama of life and human welfare; from it one sees in merciless clarity what in the coming century may overtake the rest of Africa and the world. In the rich fields along the Nile, the Ganges, and the yellow rivers of China, and in the valleys along the hillsides of Japan, in fact almost everywhere, all the lands that are really livable, all the fertile and healthy farm lands, are occupied. The nations that can afford it resort to vast drainage systems in their swamps and lowlands, to irrigation of their deserts, and to grading and terracing of their hillsides to extract for human survival the minute and few remaining morsels of livable lands, but all of these serve not to better man's lot but only to increase his numbers. Temporary surcease for a few, but mass troubles for all—unless some new force controls the destiny that lies in numbers.

The multitudes consume the crops and then, from dire necessity, destroy the land from which they get their sustenance. And yet we speak of man's conquest of nature as though it were an unmitigated blessing, and our statesmen and politicians with no recognition of population growth, carelessly speak in one breath of both eliminating disease and raising the standard of living of the world. Without victory over man's careless irresponsibility for his reproduction we can never achieve the millennium that has been so nearly within our grasp, but instead fumblingly and ignorantly, we may go down to defeat.

I had seen the animals of my native land, and I had speculated about their preservation and the fate of man. Now the date of departure was at hand. In these last days the beauty of the country seemed more impressive and poignant to me than ever. As I looked across the misty valley, the vast expanses of virginal, tender,

emerald green were maturing to lustier hues, predicting fruitful-
ness.

Every patch of bush, garden, and swamp became suddenly
irreplaceably precious. Natural beauty clamored for attention, but
only brought to mind the problems created by man and by his
behavior toward his fellow man. With a crescendo of painful
realization, I perceived more sharply than before that without
first solving the problems of man, most forms of plant and animal
life of this fair land would in due time be destroyed forever; and
that, beyond the bounds of South Africa, almost everywhere on
our globe, man's uncontrolled increase in numbers will eventually
threaten all forms, although a few adaptable species may locally
survive. This was a gloomy mood in which to end what was in
all likelihood my last visit to the land of my youth.

And yet, as I left the mission on my first stage of departure, a
Cape ringdove and a collared barbet were singing in the coral
tree shading the porch of my old home. As I started driving down
the road toward the river en route to Durban and the airport, their
fading calls were a reminder of the persistence and adaptability
and vigor of all living things.